A Handbook of Biological Investigation

Fifth Edition

Harrison W. Ambrose III
Katharine Peckham Ambrose

6507 Bull Run Rd.
Knoxville, Tennessee 37938

Hunter Textbooks Inc.

Hunter Textbooks Inc.

823 Reynolda Road
Winston-Salem, North Carolina 27104

Dedication

This book is dedicated to J. Hill Hamon and to Donald Y. Young, remarkable teachers. When I was an undergraduate student, I was fortunate enough to have J. Hill as one of my instructors. His dedication to teaching and his willingness to help the individual student not only helped me make my career choices but has always served as my example. Later, as a faculty member, struggling to introduce and test the investigatory laboratory teaching technique, I was privileged to have the help and support of another such teaching assistant, Donald Y. Young. Don too has the boundless energy, patience, and skill needed to help another generation of biology students.

H.W.A. III

ACKNOWLEDGMENTS

The authors gratefully acknowledge their debt to Elisabeth B. Davis, Biology Librarian at the University of Illinois, for her generous contribution of Chapter Nine: An introduction to Biological Literature, and for her help with Chapter Eight. We also thank Wynne Brown, a professional illustrator, for Chapter Twelve. Nicholas J. Cheper and Donald Y. Young generously helped us with Chapter Seven. In addition we would like to thank John Christy, Steve Kessell and Don Riker who, as teaching assistants, contributed their suggestions, criticisms and creative energy toward our development of the investigatory laboratory teaching technique. We would also like to thank the many other dedicated teaching assistants at Cornell University, University of Illinois and University of Tennessee at Knoxville who, like those named above, were willing to give more than was required. Finally, we thank in advance all who are kind enough to give us comments and criticisms that help us revise and improve this handbook.

CONTENTS

Much of our educational system seems designed to discourage any attempt at finding things out for oneself, but makes learning things others have found out, or think they have, the major goal. It is certainly true that it is easier to teach a set of facts than it is to encourage an inquiring mind . . . to make its own discoveries. Once a student has learned that he can find things out for himself, though, bad pedagogy is probably only an irritant.

Ann Roe in *The Making of a Scientist*

Introduction

This handbook is neither a text nor a laboratory manual. It is written as a reference aid to be used by biology students at any time in which they are first exposed to the scientific method and to original research. It guides the beginner through elementary experimental design, data collection and analysis, and the subsequent writing of a scientific report. In addition, it provides an introduction to the literature of biology and a basic search strategy for finding relevant, published information in a library. We anticipate that this book will be used in conjunction with a text in a particular subject specialty, a manual of laboratory techniques, a guide to field collection, or with a sourcebook on the care and feeding of laboratory plants and animals. None of these books alone equips the undergraduate or beginning graduate student with the potential for making his own scientific discoveries. The supplementary use of this handbook, however, can add the excitement of original student research to any course in the life sciences.

At the Universities of Tennessee and Illinois, this material has been used in sections of introductory biology which were taught in the investigatory laboratory mode — a method of teaching which prepares and later permits students to design, pursue, analyze, and report on independent research. The students were never asked to learn the material in this book but merely to use it as they would a dictionary or other reference. No mathematical ability was required; laboratories were equipped with calculators for student use.

Portions of this book have also been successfully introduced at advanced undergraduate and beginning graduate levels. No matter when a student begins original scientific experimentation, a careful, gradual introduction is required. It is the consensus of those who have taught investigatory laboratories that the enterprise is doomed if students are not adequately prepared before they begin independent work, if they are not given appropriate credit for the actual experimental effort (as well as the eventual scientific report), if the instructor insists that experiments be truly original and/or successful, if the instructor insists that the final reports be of publishable quality, and most assuredly if the instructor believes the students incapable of high-quality, creative, independent work.

For those interested in how freshmen might be exposed to independent research, we have included a general outline of the first crucial weeks in the laboratory (see the Appendix to this handbook). For more information on the concept of the investigatory laboratory, we strongly recommend the seminal report entitled "The Laboratory: A Place to Investigate" prepared by the Commission on Undergraduate Education in the Biological Sciences in 1972.

Chapter One
What Is Science?

Science is the investigation of rational concepts capable of being tested by observation and experimentation. The analytical method, which involves a pattern of observation, experimentation, and both inductive and deductive logic, is what distinguishes science from other disciplines. Biology is the science of life on earth.

The process of acquiring scientific information employs the postulating and testing of hypotheses. **Hypotheses are possible explanations for phenomena observed.** For example, if the fern in your dormitory room dies unexpectedly, you may hypothesize that God killed it. If you do, your hypothesis would not be susceptible to testing through the scientific method, although your hypothesis might be verified by revelation through prayer or meditation. This method of acquiring information lies outside the scope of science. **Science limits itself to the study of the physical universe.**

Let us suppose that you hypothesized that your plant died because it did not receive enough water. This would be a good guess because you already know that water is essential for plant life. Maybe you fear that you let a week go by without watering it and that has caused its demise. Your roommate says, however, that you did not need to water it more than once a week anyway. You are sure that water only once a week is nowhere near enough. You predict that if you bought another fern and watered it only once a week, it too would die. **Scientists use deductive (if . . . then) logic to make and then test predictions on the basis of their hypotheses.** In a sense, you have now made a prediction on the basis of your hypothesis.

Let us restate your prediction in this way: "*If* I buy another fern of the same kind, in the same kind of soil and container, and house it in the same part of my room, and *if* I water it only once a week, *then* it will die." Your roommate says, "Rubbish!" You are curious enough to go out and buy another fern. (It is the possession of a curious mind that unites scientists everywhere no matter how diverse their backgrounds, interests, and education.)

You perform your experiment and, after two weeks, your second fern dies. "I told you so! Watering only once a week is not enough," you inform your roommate.

"You must have done something different with it this time," he retorts.

"No, everything was the same as before — same kind of plant, same place in the room, same everything."

Although neither of you ever need to consciously think these thoughts, you are both making some of the basic assumptions that all scientists make. You are assuming that **there is order in the universe,** that **the human mind is capable of comprehending it,** and that, **if conditions are the same, the results will be the same.** All scientific experiments are based on the assumption that similar results would occur if the experiment were repeated under similar circumstances.

Let us look at your experiment again. Did you prove anything? You may feel that you have. Your experimental results support your hypothesis, but your roommate remains unconvinced. One of the hazards of deductive logic is that **both true and false hypotheses may give rise to true predictions.** A noteworthy example of this is the fact that the mistaken hypothesis that the sun revolved about the earth gave rise to true predictions about sunrise and sunset that were convincing enough to make that false hypothesis appear to be a law of nature.

You may now be convinced that inadequate watering has caused the death of two plants. Your roommate suggests that you watered them both too much. He is even willing to sponsor some additional research and brings home three more ferns. You hypothesize that *if* you water your plants every day (all other conditions

2

being the same), *then* they will live. You choose one plant to be kept in wet soil by watering every day, one to be watered every other day, and one (to satisfy your roommate) to be watered every 10 days. What if all the plants die? Now have you shown anything? This time you have. You have failed to support your hypothesis that inadequate watering was killing the ferns. You may be discouraged, but actually you have succeeded in rejecting your hypothesis. If you can reject an hypothesis, you have taken a step forward. **All science advances by the rejection of hypotheses. There is no such thing as proof.** Each step forward involves disproof.

You and your roommate might go on testing ferns with differing amounts of water for quite some time. If you did, you would have fallen into one of the most common traps in science. You would have fallen in love with your original hypothesis that water was somehow involved in the death of your ferns. It is a sad fact that many scientists become overly attached to the beauty and simplicity of their first hypothesis and waste a great deal of time, money, and effort testing it, even in the face of evidence to the contrary. After all, the scientist is usually very knowledgeable, so his first hypothesis is a very "educated guess." And so was yours, since water is clearly linked to the health of plants. At this point, whether you realize it or not, you have probably eliminated the possibility that water was a factor.

However, you and your roommate may still be interested in testing ferns in an attempt to determine the cause of this high mortality rate. You may well be out of money and have only your curiosity and enthusiasm left. This is the normal condition of a scientist. You may want to apply for a grant.

Why would anybody care about your specific ferns? Well, probably nobody does. However, **good scientific research produces results from which you can generalize.** You may be able to establish something about the care and feeding of ferns in general. Perhaps you approach your botany department, explaining what has happened to all your ferns, and requesting a small grant to cover the purchase of additional ferns. If they see anything interesting about the death of your ferns, they may be willing to help you, provided that you supply them with a research proposal. They will point out that water does not seem to be a factor, but have you considered that the plants may not have had enough sunlight? (In fact your plants were on a bookshelf in a rather dark corner of the room.) Or that the room temperature might have been inappropriate? (In fact, because of an energy shortage, your dormitory has been maintained at 60° F each night and it is not unreasonable to suppose that this might be too cold for a fern.) Or that, after a certain age, the ferns are too large for the pots they are sold in and need repotting? If you are ever to establish the cause of death of your plants, you will have to examine each of these possibilities — and there may well be others you have not yet thought of. Maybe somebody in your dorm is using your plants as an ashtray, for example. **Appropriate experimental design is crucial to the acquisition of scientific knowledge.**

There is a very well-established method of attack on a problem such as your. **The two essential first steps are a search in the library to find out what is already known** (there may well be a book all about the care and feeding of ferns) **and a brainstorming session.** In the latter, you and your friends should list all the possible explanations you can think of. Each explanation should be written down and turned into an hypothesis. By making predictions (if I do this, that will happen), you should be able to figure out what experiment would *disprove* each hypothesis. If you hypothesized that putting the fern in the south-facing window would allow it to live (and your reading has suggested that this is normally appropriate), then it is easy to see that putting some ferns in the window will be part of the next experiment. If these live, you have proven nothing, but you may be able to *infer* that sunlight was the problem all along. If they die, however, you can check off the too-little-sunlight hypothesis from your list. It was clearly not the factor causing death.

You should work out an experimental design in which only **one variable factor is examined at a time** while all other factors are being held constant. Perhaps three plants will be on the windowsill and three back on the dark side of the room but warmed by a space heater (with precautions that the heater not also provide light). Three other plants may be purchased and repotted, but kept on the bookshelf as before. If your literature agrees with this formula, all the plants will be watered once a week. In each experiment, the hope is to manipulate only one variable at a time. Certain experimental subjects are normally held as **controls** — in this experiment that would mean that some plants experienced no manipulation of light, heat, or pot. If you actually performed this experiment and all of your plants died, each of your variables would have been eliminated as the causative factor, all hypotheses rejected, and you would be back at the drawing board, thinking up new hypotheses. This is not an unusual occurrence. Other factors would have to be explored such as manipulations with the soil. Let us suppose, however, that you reject all your hypotheses but the too-little-sunlight one. Would you have proven that inadequate sunlight caused the death of all your other plants? Not really, but you would have grounds for a **strong inference** that sunlight was the factor. The more ferns you see living happily in south-facing windows, the stronger would be your inference. As we have said before, there is no such thing as proof. Only disproof. For anything you feel you have established, there is always the possibility of an experiment that would disprove it. (If anybody ever succeeded in raising ferns in the dark, your work and that of others would be in doubt.) Whenever you design an experiment to test an hypothesis, try to think of the experiment that would *disprove* it. If you fail to disprove it, you can infer (but never absolutely prove) that your hypothesis was correct.

Many rather subtle points have been touched on in this introduction. Each will be amplified and clarified during the course of the book. We hope that this hypothetical situation has shown you several things. First of all, that any unusual happening which is not immediately explicable (such as the sudden death of a fern) might stimulate a small-scale scientific experiment. Experiments frequently lead to new questions. Scientists are often led from question to question, gradually checking off rejected hypotheses, as if they were detectives solving a mystery. In fact, a scientist is a detective, and nature is filled with mysteries. Adequate reading and preparation can save a great deal of time and effort, but only a very careful plan of attack, in which variable factors are segregated and examined one at a time, can ever hope to show anything valuable. (There are some experimental designs and statistical analyses that allow one to manipulate more than one variable at a time, but these will be saved for later.) However, even a minor experiment, which provides no new information to add to the heaps of accumulated scientific data, can involve the challenge, the mental exercise, and the fun of "big science."

Chapter Two
Asking Questions

S cientific information is gathered by the asking and answering of questions. Not all questions are susceptible to scientific inquiry, as we have said before. Questions which do not deal with the physical world (What is truth? Is it right to "mercy kill" a person suffering from a painful, terminal disease?) are outside the scope of this discipline. Extremely broad questions about physical phenomena are also inappropriate for our purposes even though they fall within the scope of natural science. For example, asking "What is sunlight?" would be too broad a question, but asking whether sunlight were required for the germination of a seed would be entirely appropriate. Or again, asking "What is gravity?" would be too broad for the beginner to get an experimental handle on it, but a question about the effects of gravity, on plant growth, for example, would be entirely suitable for study.

Before you are ready to select a topic for scientific research, you must practice asking questions about the physical world and figuring out what kind of information would be needed to answer these questions. Those of us fortunate enough not to have had our curiosity stifled by exasperated parents or teachers are constantly asking ourselves questions. We wonder when the robins will return, whether or not the coat of the woolly bear caterpillar really can be used to predict the severity of winter weather, and why the male cardinal is more brightly colored than his mate. If, on the other hand, our childhood questions were frequently answered with the uninformative, "Because that's the way things are," we may have lost our natural curiosity and we may find that we think of no questions at all. To rekindle the spark, we must now exercise our minds, consciously looking for questions and trying to think of ways to answer them.

Let us start by asking the very simple question, "How many students are in my biology laboratory section?" Obviously, all we must do to answer this question is make a simple head count. If we had asked a question of a comparative nature such as whether there are more men than women in the class, the data collection would be slightly more complex. In this case the population under study (the class) would have to be categorized into two groups (men and women), each group would have to be counted, and the totals compared. You will note that the term **population** was used to describe the subjects of the latter study. Most things come in populations, or groups of similar kinds of things. You could use the term when discussing cars, birds, plants, numbers, or people. No matter what population of items you are studying, the kind of question you are asking determines the kind of data you collect. Although it was perfectly obvious in the two sample questions about the students in your class, it must be stressed that, **in order for any test to measure the property being considered, the data collected must be relevant to the question.** There are different kinds of data and different scales of measurement for answering questions about them.

Kinds of Data and Scales of Measurement

When we identify or estimate various **parameters** (aspects) of a **population** (a group of similar kinds of things), we may use one of four different scales of measurement. These scales of measurement differ in their power to provide us with information and in the degree to which they supply us with data susceptible to statistical analysis.

The first two scales of measurement, the **nominal scale** and the **ordinal** (or **ranking**) **scale**, are appropriate for **discrete data**. In discrete data, each item is a separate, whole unit. Cats and dogs are discrete units because there is no possibility of a cat-and-a-half falling between cat and dog. The head count of members of your class was a measurement of discrete data.

When discrete data are measured on a **nominal scale,** they are merely collected and grouped by arbitrary names, numbers, or symbols. A population of cars could be divided into groups named Dodge, Chevy, Ford, etc. People could be grouped by sex into the categories male and female. The classifications of nominal data are equally weighted. The category names (such as Dodge or Chevy) have no particular relationship to each other except that they describe the same population. Even when numbers are used to describe categories of nominal, discrete data, the numbers are arbitrary — rather like the numerals on the shirts of football players. (Football numerals could be considered equally weighted because the player with the highest number is not necessarily the heaviest, fastest, or best; the numbers are merely identification.) **Nominal measurement is merely a sorting of a population into named groups;** it is the weakest form of measurement. Nominal measurement can be used to answer questions of how many and how frequent. You might identify the butterflies in your backyard by their species name and count how many you find of each type. This would be a measurement of discrete data on a nominal scale. The data would only tell you how many you found in each category and which species occurred most frequently. More complex questions require a more powerful scale of measurement.

Some nominal, discrete data can be organized along an ordinal or ranking scale. **When a population is measured on an ordinal scale, the named categories are organized in terms of some relationship they have to each other.** Order can be imposed on the nominal data about cars by ranking the categories by weight, for example. VWs might be the lightest category, followed by Chevy, Dodge, Buick, etc. Note that the items, although organized into a relative order, are still discrete. There will

9

be no Chevy-and-a-half on the scale. Nominal categories such as military ranks (sergeant, major, etc.) are also examples of categories that may be ordered on a scale of relative value or importance. When things are measured on an ordinal, or ranking, scale, the resulting data, besides yielding information about how many and how frequent, can also give you information about a central tendency called the median. (The median, which will be discussed more fully in Chapter Three, is the middlemost score or rank in a series of ordinal measurements.) If you are doing an experiment in which you are testing an hypothesis about a correlation between the discrete data you have collected on an ordinal scale and some other phenomenon (such as a test to see whether there is any preference for cars in the heavier categories), there are ranking statistical tests which apply. (The Spearman's Rank Correlation test in this handbook is an example.)

The first two scales of measurement we have discussed, the nominal scale (which is basically a naming of categories) and the ordinal scale (in which named categories are ranked in some order), are both suitable for discrete data in which each item is a separate, whole unit. More information can be obtained by measuring on an **interval** or a **ratio scale.** Both of these scales of measurement are for **continuous data.** Continuous data are points taken along a scale that, at least theoretically, could be subdivided. If you were noting times of a particular phenomenon and your readings were 7:00, 7:30, 8:00, 8:30 . . ., there always would be the possibility of additional data points such as 7:31, 7:45, etc. Unlike discrete data, which involve whole units such as cats or cars, **continuous data are taken from some continuum such as time, weight, or temperature.** Additional data points between those observed and recorded are always theoretically possible. As we have said above, there are two scales of measurement for continuous data, the interval scale and the ratio scale. Unlike the ranking scale, which merely gave an order to named categories,

the interval and the ratio scale both indicate the distance between the items. The intervals, or distances between the categories, are purely arbitrary and the ratio between any two intervals is independent of the unit of measurement and of the zero point. (We will illustrate this in a minute.) The only difference between the interval scale and the ratio scale is that, in the interval scale, there is no actual, "real world," zero point.

The most common example of an interval scale of measurement is the measurement of temperature. Temperature is customarily measured on either one of two different interval scales called Celsius and Fahrenheit. Because there is no such thing as a complete absence of temperature, there is no "real world" zero point. In each of these interval scales an arbitrary zero has been established (based on the temperature at which water freezes in the case of the Celsius scale).

To illustrate, let us assume that these temperature readings represent some data collected in both Celsius and Fahrenheit. Notice that these two interval scales and their zero points are arbitrary but that the intervals, or gaps, between readings are proportional.

CELSIUS	0	10	30	100
FAHRENHEIT	32	50	86	212

If you do not see that the data are proportional, note that the size of the gap between 0° and 10° (10) and between 10° and 30° (20) is the same ratio (10:20) as that of the corresponding gaps in the Fahrenheit readings (18:36). In other words, the differences between temperature readings on one scale are proportional to the equivalent differences on the other scale but the numbers themselves are arbitrary.

It is theoretically possible to create an interval scale using non-numerical categories since the unit of measurement is arbitrary, but before any meaningful statistical manipulation could be performed, it would be necessary to establish that the intervals or distances between the categories were equivalent. In fact, it would be necessary to translate them into numbers before any arithmetic operations could be performed.

Data measured on an interval scale are susceptible to more statistical treatment than are nominal or ordinal data. Because the interval scale is truly quantitative, many statistical parameters (aspects) may be estimated. Interval scale measurements can yield information about means, standard deviations, and correlations (defined in Chapter Three) and can be evaluated by all the common parametric statistical tests.

The other powerful, quantitative, and statistically useful scale of measurement for continuous data is the ratio scale. **Measurements on a ratio scale differ from those on an interval scale only in the respect that they always have a true zero point.** Once again the actual units of measurement (inches, millimeters, etc.) are arbitrary. In the following example of continuous data measured on a ratio scale, note that the measurements in inches and in millimeters share a common zero point. And beyond that point, the units of measurement are arbitrary and, again, proportional.

INCHES	0	1	10	20
MILLIMETERS	0	25.4	254	508

It is important to bear in mind that the kind of data you collect in the experiments you design will dictate the way the data can be analyzed statistically. It is important to consider from the start what you will measure and how you will treat your data. If you are merely grouping cars by kind to see how many of each kind go by the window, you are collecting nominal, discrete data. If you are interested in whether more people drive heavy cars, you may have ranked your nominal categories on an ordinal scale by relative weight. If you are stopping each passing car and driving it up on a scale to record actual weights, you are collecting continuous data which will be measured on a ratio scale. If you find later that you need discrete data, but you are still interested in using the weights you have recorded, you may wish to distribute your data into some arbitrary, but discrete, units of measurement such as 0 to <100 lbs., 100 to <200 lbs., 200 to <300 lbs., etc. It would now be converted into a discrete, ordinal scale. Understanding what kind of data you need to answer a particular question, and being able to show statistically that your results were not merely the result of chance, involve advance planning and an understanding of the concepts presented in this chapter.

Chapter Three
Description of Data:
Central Tendencies

You will need a way of interpreting whatever kind of data you gather. If you have taken a random sample of a population, estimating one or more parameters (aspects), you now have a series of readings. Let us suppose that you have measured the height of each tree on a previously unexplored island. Actual tree heights are continuous data, measured on a ratio scale, but to simplify your experiment, you have limited your measurements to discrete units of one meter each.

One of the most useful and illuminating first steps in organizing data is to plot it on a graph so that it can be visualized more easily. For many types of experiments, data can be plotted on a **frequency histogram.** Traditionally, the vertical axis (called the **ordinate**) stands for the frequency of occurrence of a particular measurement. The horizontal axis (called the **abscissa**) always represents your chosen units of measurement. The lowest measurement is always plotted at the extreme left; the measurements get progressively higher as you proceed to the right. Figure 3.1 is an example of a frequency histogram for recording tree heights measured in one meter units.

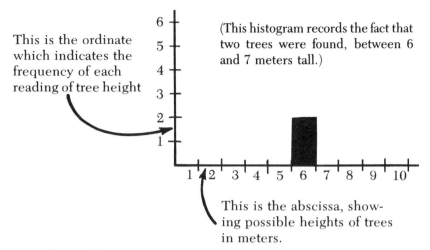

This is the ordinate which indicates the frequency of each reading of tree height

(This histogram records the fact that two trees were found, between 6 and 7 meters tall.)

This is the abscissa, showing possible heights of trees in meters.

Figure 3.1

Once you have plotted your data, it may be obvious that it is distributed with some particular pattern of frequency. It is possible, but unlikely, that your data will be uniformly distributed. A **uniform distribution** of data would look like this.

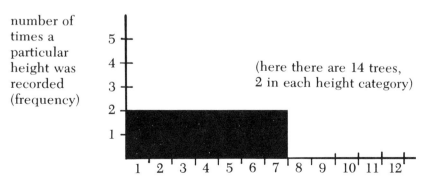

number of times a particular height was recorded (frequency)

(here there are 14 trees, 2 in each height category)

Figure 3.2

heights of trees in meters

It is important to understand that if the data looked like Figure 3.2, the trees did not look like this:

Figure 3.3   This is a measure of height in meters.

If the woods had looked as we have pictured it in Figure 3.3, the data points would have fallen into a single, tall column over the two meter mark on the abscissa.

The uniform distribution of tree heights plotted in the histogram (Figure 3.2) might have come from a woods that looked like Figure 3.4 (there are an equal number of trees in each category).

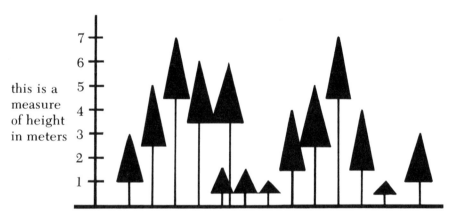

this is a measure of height in meters

Figure 3.4

It is also possible that the data you plot might show a **skewed** distribution. In a skewed distribution, the data reveal an obvious central tendency but they are not

16

evenly distributed on either side of the high point. Data can be skewed to the right, in what is called a positive skew, or to the left, in a negative skew. In our example of skewed data (Figure 3.5), we see a negative skew; the low readings are more extreme than the high ones.

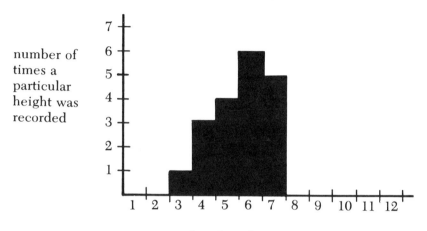

number of times a particular height was recorded

heights of trees in meters

Figure 3.5

The trees represented on Figure 3.5 might have looked like those in Figure 3.6 — note that there are many trees in the taller categories.

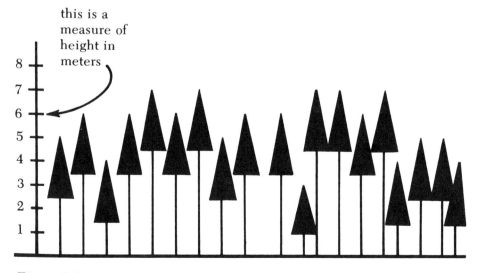

this is a measure of height in meters

Figure 3.6

Another kind of distribution which your data might display when plotted on a frequency histogram is called a **bimodal** distribution. In a **bimodal** distribution the data seem to fall into two groups. In Figure 3.7 we have plotted another histogram of tree heights — this time with a **bimodal** distribution.

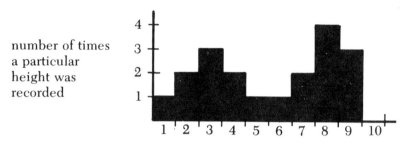

number of times a particular height was recorded

heights of trees in meters

Figure 3.7

Often when your data falls into a bimodal distribution, there is reason to suspect that the population you have been studying is really two populations. The woods represented in the bimodal distribution in Figure 3.7 might have looked like this:

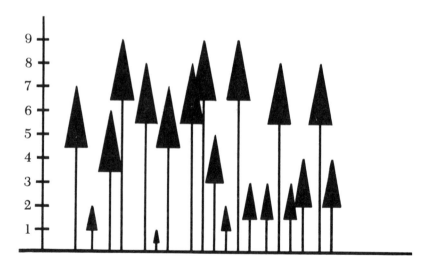

Figure 3.8

Or it might have looked like this:

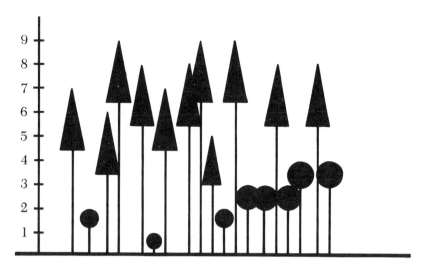

Figure 3.9

Perhaps you were treating two different species as if they were one. Even when you are clearly dealing with only one species, such as *Homo sapiens* (man), you might plot your readings of adult body weights and find that they fall into a bimodal distribution. In this case your readings would indicate two central tendencies, one for men and one for women.

There are other ways that data can be distributed; there are random distributions, log normal distributions, etc., but the most frequent kind of distribution found for a sample of a single population is called a **normal distribution**. It is equally spread out on either side of a central high point with a characteristic "bell" shape. Figure 3.10 is an example of a normal distribution of our trees.

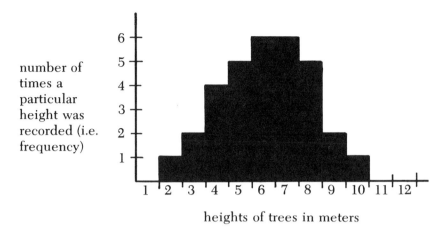

number of times a particular height was recorded (i.e. frequency)

heights of trees in meters

Figure 3.10

Most data, even if somewhat skewed, are spread out on either side of a central high point. When you analyze your data, you are often interested in reporting what usually happens — in other words, the central tendency. There are three common measurements of central tendency — the mean, the median and the mode.

The **mean**, or numerical average (often written as $\bar{X}$), is the simplest measurement to make. All you have to do is add up your various measurements and divide this total by the number of measurements taken. To calculate the average, or mean height of the trees on your study island, you would add up all the different heights and divide this sum by the number of trees in the sample measured. The mean is useful in many statistical tests, but, reported alone, may not be very meaningful. If your plotted data were very skewed or bimodal, the mean might be deceptive. A mean is always strongly affected by any extreme readings.

Another measurement of central tendency, one that is less susceptible to distortion by an extreme reading, is the **median.** Before you can calculate the median, you must organize your readings in a progressive sequence. If you were going to calculate your median tree height, you would put all your measurements in order with the shortest first and the tallest last. The median of the distribution of data is the middlemost value. When you have an even number of readings, the median falls between the central figures. You can calculate an exact median by averaging the two central data points (adding them together and dividing by two).

Once you have your data organized sequentially, you can easily find the **mode.** The mode is the most frequently occurring value. There can be more than one mode to a distribution as we have seen in the example of the bimodal distribution.

If your data fit a *perfect*, normal distribution, the mode, the median, and the mean would be the same figure, *by definition*. However, in many real distributions of data, that are called "normal" (because for statistical purposes they do not differ significantly from the theoretical normal), these three measures of central tendency need not be the same, and often they are not. If your measurements were the following: 2, 3, 4, 4, 5, 6, 8, 10, and 12, the following would be your measurements of central tendency:

the **mean** would be 6 (the sum of the values, 54, divided by the number of values, 9);

the **median** would be 5 (the middlemost value); and

the **mode** would be 4 (the most frequently occurring value).

Chapter Four
Description of Data:
Dispersion

We have indicated some of the ways to measure a central tendency in your data. Often it is useful to know what is happening outside the central tendency — to have some measurement of the dispersion or spread of the data about the mean ($\bar{X}$). Dispersion can be measured in terms of the **range** of the data; this is merely the "distance" between the lowest and the highest readings.

Let us assume that you have measured the height (in meters) of seven trees, selected at random, from each of four different islands. Your data, when organized from the lowest to the highest reading might look like this:

Island I	8	8	9	10	11	12	12
Island II	5	6	8	10	12	14	15
Island III	1	2	5	10	15	18	19
Island IV	8	10	10	10	10	10	12

Figure 4.1

Note that each island had the same average tree height. If the means were the only data you reported, the islands might seem remarkably similar. In fact, there was a considerable variation between them. Island I and

IV had a range of 4 meters, and Island III had a range of 18. In this instance, calculating the range of the data might be quite important. Here is a display of the data from our four islands reporting the means and the ranges of tree height in meters.

Island I	$\bar{X}$ = 10 M	range = 4 M
Island II	$\bar{X}$ = 10 M	range = 10 M
Island III	$\bar{X}$ = 10 M	range = 18 M
Island IV	$\bar{X}$ = 10 M	range = 4 M

Figure 4.2

However, reporting only the mean and the range does not tell anything about how clustered the readings were. The bulk of your data might not deviate much from the mean. It is nice to know how many readings were close to the mean and how many were widespread. The other measurement of dispersion, or spread in data, is called the **variance.** It is a statistically more useful measurement which conveys more information than the average or the range alone. In order to understand the concept of variance, it is necessary to know more of the mathematical properties of the normal distribution. (The formula for calculating the variance will be given on page 29.)

Although your data will never exactly coincide with the perfect, normal distribution, it may be fairly close. It will usually be appropriate, therefore, to describe it in the terms we will give for a normal distribution.

The data in Figure 4.3 are normally distributed. Again the frequency is plotted on the ordinate, and the unit of measurement (whatever it may be) is plotted on the abscissa.

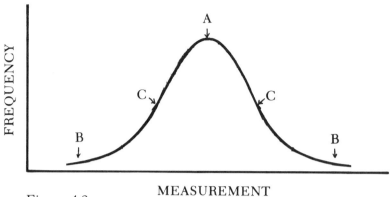

Figure 4.3   MEASUREMENT

You will note that the step-like characteristics of the histogram have been removed because, in this plotting of data, all the theoretical points between each step are represented, smoothing the steps into a line. (Data are often displayed with a curved line even when not all the intermediate steps have been recorded.) On this curve, often described as a bell-shaped curve, the highest point (A) is the mean ($\bar{X}$), the median, and the mode. The two tails of the curve (B) are not drawn touching the baseline. The two lines (tail B and the abscissa) are said to be **asymptotic** to each other. They are presumed to recede indefinitely, coming closer and closer without ever touching.

There is another mathematically and theoretically important feature to the normal curve, the **points of inflection.** The points of inflection (C) are points at which the curve changes from convex to concave. For the purposes of demonstrating the variance, theoretical lines can be drawn to the abscissa from the two points of inflection and from the mean point. The distance between the mean point on the baseline and the point of inflection, on the baseline, is called one **standard deviation.** It is a sort of average of the deviation from the mean. The symbol for the standard deviation is "s".

Here is another normal curve with the standard deviations drawn.

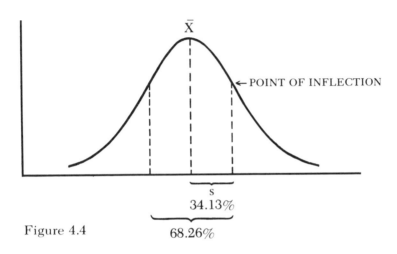

Figure 4.4

Of the data you collect from a normally distributed population, 34.13% *by definition*, falls within one standard deviation from the mean. One standard deviation on either side of the mean includes 68.26% of your data.

It is very helpful to report both the range and the standard deviation of your data. A particularly clear way to summarize data taken from more than one population, without having to plot all the different readings, looks like Figure 4.5. Here we have taken the data from Figure 4.1 for the trees on the four islands and shown the means, the ranges, and the standard deviations. (Note that there is no abscissa — the vertical scale gives tree heights in units of one meter.)

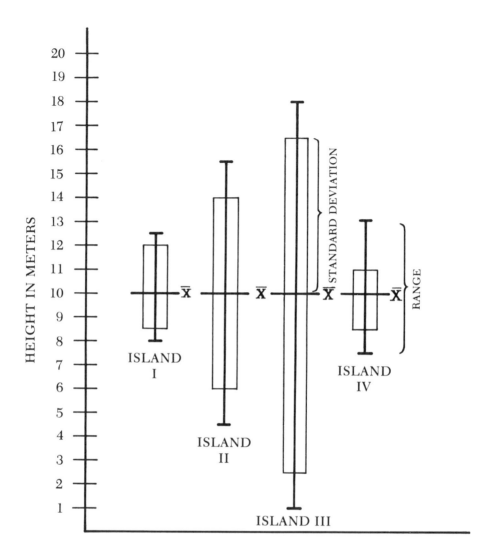

Figure 4.5

In the display of comparative data in Figure 4.5, the fact that the average height of each population is the same is clearly displayed. In addition, the extreme variation in the range of the data is obvious. The boxes which show the standard deviation on either side of the mean help the reader understand how clustered or dispersed the data were (68.26% of the readings fell inside the boxes).

Sometimes it is useful to talk about "two standard deviations from the mean" or even three. The standard deviation is an arbitrary, fixed length on the abscissa. If you measure out two standard deviations from the mean in each direction, you have accounted for 95.44% of the data. Three standard deviations would include 99.74% of the data. Any reading which fell outside three standard deviations from the mean would be an extreme one indeed. However, there always is the probability of such a reading.

Here is the normal curve again, with three standard deviations drawn on either side of the mean.

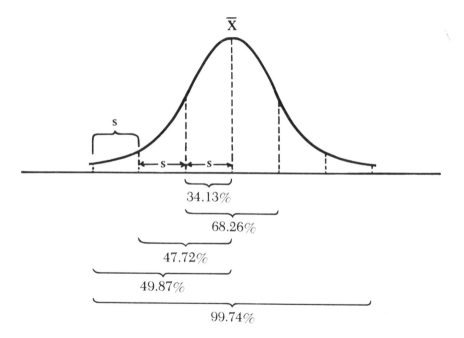

Figure 4.6

Because the standard deviation is a particularly useful and important aspect of your data, it is helpful to have an easy means to calculate it. It is not done by dropping lines from the inflection points on the curve of your data because the inflection point is a theoretical point, and your data will not fit the true, hypothetical normal curve.

There is a mathematical formula for calculating "s" — the standard deviation. The formula is appropriate for any data that are fairly close to being normally distributed. (If you are doubtful about your data, even after you have tried plotting them on a histogram, there are mathematical tests for evaluating the distribution of your readings.) Although the formula looks complicated at first glance, you can use it whether or not you ever come to understand it. The data are merely inserted according to the directions. (In this handbook, none of the mathematical calculations are difficult ones and previous mathematical training, although useful for the ultimate understanding of such formulas, is not required.) Some calculators are programmed to calculate the standard deviation for you.

This is the formula for calculating the standard deviation of your data.

$$s \cdot = \sqrt{\frac{\Sigma x^2 - \frac{(\Sigma x)^2}{n}}{n - 1}}$$

s = the standard deviation (this is the figure you are going to calculate, you do not plug it in at the start)

Σ = the mathematical symbol for sum

x = any one of your data points

x^2 = x times x

Σx^2 = square each of your data points, then sum these squares

$(\Sigma x)^2$ = the square of the sum of all your data points (add up the data points and then square the sum)

n = the number of data points you have collected

n − 1 = one less than the number of points collected

In other words, 1) square the values for each of your data points and then sum these squares, Σx^2; 2) sum the values for each of your data points and then square this sum $(\Sigma x)^2$; 3) divide the value obtained in step 2 by the number of data points, $\dfrac{(\Sigma x)^2}{n}$; 4) subtract the value obtained in step 3 from that obtained in step 1; 5) divide the result of step 4 by the number of data points minus 1; and finally, 6) take the square root of the figure you have calculated. This is the standard deviation of your data.

If you now look back at the display of the data on tree heights in Figure 4.5, you will see that in populations with a high degree of similarity, the standard deviation is small. In highly dissimilar populations, the standard deviation is larger. This is why it is such a valuable measure of the dispersion of data.

As we have indicated earlier, there is one other frequently used measurement for describing the dispersion of data, the **variance**. The variance is defined as the square of the standard deviation, or s^2. The formula for calculating the variance is the same as the calculation of the standard deviation except that there is no need to take the square root.

$$s^2 = \frac{\Sigma x^2 - \dfrac{(\Sigma x)^2}{n}}{n - 1}$$

Because the square root is not taken, the variance is usually too large to plot in the kind of data display we used for the four tree populations in Figure 4.5. The variance is a figure that you may need to calculate for certain statistical tests. Some tests use the variance and some use the standard deviation.

In the next few chapters we will introduce the idea of the null hypothesis and then proceed to the analysis of data in order to test the null hypothesis.

Chapter Five
The Experimental Plan

One of the most difficult tasks for the beginner is the selection of a research topic. It is absolutely essential that the initial questions chosen can be answered through simple observation or experimental manipulation, using relatively unsophisticated techniques, in a reasonable amount of time. If you are in a research course, a general menu of available species and equipment will probably be supplied. If you are on your own, you might wish to examine some books about research problems and laboratory techniques (a few are mentioned in the bibliography at the back of this handbook). Or, you can simply find a topic by gradually narrowing the field of possibilities until you hit upon something feasible. You might ask yourself first whether you are interested in plants or animals. If animals, large or small? Unless you live on a farm, in a forest, or at the zoo, it is wise to select a relatively small and preferably inexpensive species. Again, your geographical situation should help you choose between terrestrial or aquatic animals, etc. Be guided by the practical aspects — what is actually available to you in nature, in somebody's laboratory, at the pet shop, or in the catalog of a biological supply house. Once you have focussed on a handy species, say some local species of ant, read about it, talk to people, and see, for example, if you are interested in its anatomy, physiology, or possibly its social behavior. Think about how you would study this species. Would you work out of doors? In your apartment? Would you need to construct some housing such as glass-covered trays? These decisions will depend on the specific question you decide to ask. Eventually, you must focus your attention on a very small aspect of the ant's repertory, for example, and ask something you can actually test, such as, "How long does the scent marking an ant's trail last?"

In most biological experiments there are numerous sources of error. To turn to another example for a minute, let us suppose that you are studying the effect of light on the amount of CO_2 taken up from an aquatic medium by an underwater pond plant. You may be using, or misusing, an electronic CO_2 probe to take readings of the amount of CO_2 in the water. You may also be measuring the light intensity with a light meter which may or may not be working properly. And your experimental plant may be in poor health without your realizing it. There may also be some problem with the logic of your experimental design, and you may be subjecting your data to an inappropriate statistical test. In any case, if, for some reason, your results are unintelligible, you will not know whether the fault lay in your measurement (either a human or mechanical failure) of the two parameters (light and CO_2), in the biological system being examined (the plant), or in your experimental plan of attack.

Therefore, we will concentrate here on the very basic elements of experimental planning. Chapter Eight, which outlines a search strategy for getting information out of a library, will help you find information relevant to your own, particular topic of interest.

I. The **first step** in planning an experiment is choosing an answerable question. Once an area of interest has been identified (I'm interested in the people passing by the door of this lab), you must focus on a specific question (Are there more men than women passing the door?). It may be your private hypothesis that there are more men than women; often an experimenter has something in mind that he expects to show. The trick in good experimental design is to capture the phenomenon in a logical box.

You will recall from the first chapter that all progress is made by the rejection of hypotheses. There is no such thing as proof. All the possible alternative explanations for a phenomenon or all experimental outcomes must be listed and an attempt made to eliminate each possibility. (In our example in the first chapter, all explanations except the too-little-sunshine hypothesis were rejected, permitting us to *infer* that the lack of adequate sunshine was the factor causing the death of the ferns.)

Because the logical experimental framework is based on the rejection of hypotheses, experiments usually begin with the statement of what is called a **null hypothesis** (abbreviated H_0). **The null hypothesis is an hypothesis of no difference.** If we are interested in whether or not there are more men than women passing the laboratory door, our null hypothesis is,

> H_0: There is no difference between the number of men and the number of women passing the door. (For statistical purposes the numbers are the same.)

Naturally you will be collecting data on the number of people of each sex passing the door. You may count more men than women and consider the experiment over. The absolute values you have counted may be different but this difference may be so slight that it is *statistically meaningless.* In other words, although you may have counted more men, chance alone could have caused this difference and it may not tell you anything important about the population under study. As a scientist, you should apply a statistical test to your data to see if you have a statistically significant reason to reject your null hypothesis. If you fail to reject your null hypothesis, your experiment is over and you have shown that, even though your absolute values may have differed, there is no meaningful difference between the number of men and the number of women passing the door. However, if you statistical test has shown that there *is* a significant difference between the two absolute values, you may reject your null hypothesis. This is an important step forward.

If you had only been interested in showing that there *was* a difference, your first (and, in this case only) alternative hypothesis would have read,

H_a (alternative hypothesis): There is a difference between the number of men and the number of women passing the door.

Normally, as you plan your experiment, you reduce your plan to a symbolic shorthand format (because it will then be easier to substitute the actual data collected for parts of the formula). The logical framework for your experiment might look like this:

H_0: # men = # women   (# means "number of ")
H_a: # men ≠ # women   (≠ means "not equal to")

However, it is likely that you would want all the possible alternative answers as part of your logical framework — one alternative (H_1) stating that there were more men and the other (H_2) that there were more women. Your experimental plan would have looked like this when reduced to its logical format:

Question: Are there more men than women passing the door?

H_0: # men = # women   (The symbol > means
                                "greater than," and
H_1: # men > # women   the symbol < means
                                "less than.")
H_2: # men < # women

If the outcome of your statistical test permits you to reject your null hypothesis, and if you have listed all the possible alternatives (in this case it is easy because there are obviously only two and they are clearly mutually exclusive, but in the fern experiment in Chapter One there were many possible alternative hypotheses), you are now in a position to pick one of the alternatives. In some instances, further experimentation may be involved before you can distinguish between alternatives, but under most circumstances, if all logical possibilities have been covered, all you need to do to select the correct alternative is to examine your data. In our people-passing-the-door experiment, once we have rejected the null hypothesis, if we have counted more men than women passing the door, we now have a statistically significant reason for concluding that more men than women pass the door. (The difference in absolute values, the actual counted numbers, was probably *not* a chance event.) (You will see later in the book that certain statistical tests such as the Kruskal-Wallis Test, the Friedman Test, and the Analysis of Variance permit you to reject your null hypothesis but do not, in themselves, allow you to select the correct alternative hypothesis.)

To summarize the first step in planning an experiment, then, you must begin by establishing your logical framework. Your experimental question must be stated in terms of a null hypothesis and all logical alternative hypotheses must be listed. You should then be able to reduce these verbal statements to the simple, logical shorthand format shown in the example above. Once the logic of an experiment has been determined, you may proceed to the second step.

II. The **second step** in an experimental design is figuring out what kind of data would be needed to answer your question. Here the concepts of discrete and continuous data become relevant. Besides knowing **what kind of data** you will be collecting, you should know **how much data to collect.** The question of how much data is more important than you might think. Of course, time can be wasted by collecting too much data, but the real danger is that you may not collect enough. Consider this, for a moment. If you were flipping a coin and, in 10 tosses, you got six heads and four tails, you would have no strong reason to suspect that the coin was defective or unfairly weighted. However, if in 1000 tosses, 600 were heads and 400 tails, you would have reason to be suspicious. And it was the fact that the 6:4 ratio was the result of so many tosses that caused the alarm. We will go into this more fully later, but the example is intended to show that, in terms of common sense, a large sample size is more convincing and meaningful in mathematical terms. The tables which accompany the statistical tests in Chapter Seven will be useful in determining appropriate sample sizes for your data.

All of the statistical tests we will give you are merely ways to examine different kinds of data, collected in different ways, to determine whether or not you have a statistically significant reason to reject your null hypothesis.

The important part of step two, in which the plan of attack is outlined, is that you know ahead of time how you will test your data. The test to be used may require a certain kind of data (ordinal, perhaps) and certain minimum sample sizes (in other words the minimum amount of data you must collect). The entire experiment should be thought through at the start.

Whether you do this as a means of organizing your thoughts, or to have a research topic approved by your instructor, or because you are applying for financial aid, it is important that you write out your research plans in the form of a **research proposal.**

The research proposal should contain a clear description of the areas of interest and of the specific question being asked experimentally. The explicit statement of the null and alternative hypotheses is also useful. In a full written proposal, there should also be a justification of the effort (and perhaps expense) entailed. This usually is an indication of why the question is important and how it fits into a broader picture of what is already known about the topic. Normally a full search of the relevant published literature precedes a serious experiment. (Chapter Eight deals with a search strategy for finding such information in a library.)

The proposal should outline the actual plan of experimental attack, mentioning specific equipment and techniques to be used (and referring to the literature if previously published procedures are to be followed). The organization of the controls and variables should be described, as well as the sample size envisioned. For students, an actual day-by-day schedule of events and procedures is particularly useful. If the experiment involves several steps, a verbalized or diagrammed flow chart of decision points is helpful as well. ("If A happens, we will do such-and-such. If B, we proceed to . . .")

Finally, the proposal should mention the statistical treatment that will be applied to the data collected. (Chapter Six will help with the selection of the right test for your kind and amount of data.) It is only with this type of careful planning that you can avoid collecting meaningless, inconclusive, or irrelevant data.

Later, when you are writing the report of your experiment (see Chapter Ten), you will find that most of the material in your proposal will be useful. The justification may be used in writing the introduction and possibly the conclusion; the description of methods and materials will be practically written; and the results section will follow logically from your statistical plan. The time spent on the research proposal will be well invested, insuring a logical experiment and an easily-written report.

We recommend the research proposal format on the following page.

TITLE:

(For help in writing a descriptive title, see Chapter Ten.)

QUESTION:

WHY THIS QUESTION:

EXPERIMENTAL LOGIC:

H_0 (null hypothesis):

H_1 (first alternative):

H_2 (second alternative):

(there may be more alternatives).

METHODS AND MATERIALS:

PROPOSED STATISTICAL TREATMENT:

(Sometimes it is useful to make up some hypothetical data of the type you expect to collect and actually try out the statistical test. This may reveal how much data would make an adequate sample.)

Chapter Six
The Statistical Analysis

In Chapter Seven you will find your arsenal of statistical tests. It is possible that you will require others not given here, but this selection will cover most of the beginning experimental situations. The collection begins with a flow chart (Figure 7.1, page 50) to aid in the choice of the appropriate test. The flow chart guides you through a series of decisions which depend on the kind of statistical question you are asking and the kind of data you are collecting. Each test is preceded by precautions which limit the test to certain kinds of data. Each test is followed by a sample experimental situation in which the test is used appropriately.

Learn to use these tests as you would use a reference book — without memorizing the contents. You need only know how to select the correct test and plug in your own data.

Because all of the tests are aimed at determining whether or not you have a statistically significant cause to reject your null hypothesis, they are all asking either one of two general kinds of questions — questions of correlations or questions of differences. On the flow chart, the first decision you must make is: Which kind of questions are you asking your data?

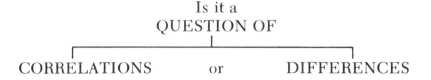

Is it a
QUESTION OF

CORRELATIONS   or   DIFFERENCES

Questions of Correlations

This sort of statistical question is often asked when you have examined two parameters of a population (such as the height and weight of a group of people) and you want to know whether there is a meaningful **correlation** between these factors. And if they are related, is there a positive correlation (taller people are heavier) or a negative correlation (taller people are lighter). When you are looking for a correlation between two factors, the data are plotted differently from the frequency histogram we have shown you. In this instance, one factor, such as height, is measured on the ordinate and the other factor, weight, on the abscissa. Each measurement you have taken would be recorded as a dot on the graph over the correct weight at the level of the appropriate height.

Data which are positively correlated might look like this:

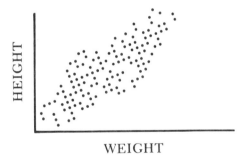

Negatively correlated data might look like this:

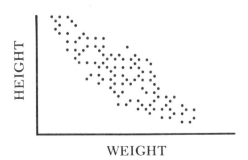

Data which are not correlated would look like this:

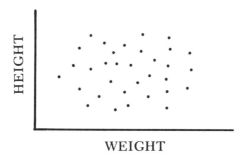

Often, of course, the data are nowhere near as obvious. That is why a statistical test is needed to permit you to reject your null hypothesis which, in questions of correlations, would read: There is no correlation between . . . (height and weight, in our example).

When two factors are correlated, the magnitude of one changes with the magnitude of the other, but no cause and effect relationship need exist. Height does not *cause* weight nor weight cause height, although the factors are clearly related.

In some situations, however, there may be a dependent relationship between two variables. The magnitude of one variable may be dependent on the magnitude of the other. This is often true in the case of human blood pressure. A person's blood pressure (the dependent variable) may be a function of his age (the independent variable), although his age would not be a function of his blood pressure. One variable depends, to some extent, on the other. **Regression** analysis is used to study the strength of this kind of relationship between two variables. Regression analysis may also be used to determine the extent to which one variable, such as SAT (Scholastic Aptitude Test) scores, can be used to predict values of another related variable, such as college grade point average.

Questions of Differences

At your first decision point on the flow chart, you needed to know whether you were testing for correlations or differences. If you are testing for differences, you must choose your statistical analysis from one of three categories — differences between distributions of discrete data, differences between means (of two or more samples), and differences between variances. We will consider each of these categories briefly.

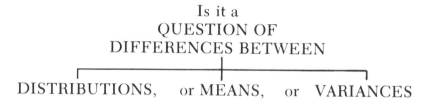

A test for **differences between distributions** of discrete data would tell you whether some data distribution differed either from some theoretical distribution or from some other actual distribution. In the first instance you might be testing the heads:tails ratio from a coin toss (say 55:45) to see if it were statistically different from the theoretically expected ratio of 50:50. Your null hypothesis in this instance would be:

H_0 : 45 : 55 = 50 : 50

If you were looking for differences between two actual distributions of discrete data, you would be testing for what is termed "independence between two samples." You might be comparing data collected on the ratio of men to women living in Town A (5,467 men in Town A : 5,688 women in Town A) to the ratio found in Town B (8,975 men in Town B : 9, 230 women in Town B). For testing independence between two samples using this example, your null hypothesis would be:

H_0 : # Town A men : # Town A women =
Town B men : # Town B women
or
H_0 : 5,467 : 5,688 = 8,975 : 9,230

Tests for **differences between means** ask the data if the mean of one population distribution is significantly different from the mean of another population distribution. This could be asked for two populations which you suspect might be different or, when you have separated one population into a control group and an experimental group, and you need to know whether your experimental manipulation has created a meaningful difference in their means. Let us say that you plotted the heights of a control group of plants (A) and of an experimental group (B), which differ only in that they have been treated with a growth hormone. Your null hypothesis is this: $H_0 : \bar{X}A = \bar{X}B$. In effect you are asking which of these two distributions represents the results of your experiment?

This?   or   This?

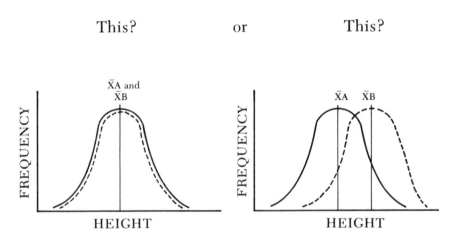

Often tests which compare the means of two populations are making the assumption that the variances (s^2 — the square of the standard deviation) are similar. You may need a test to determine whether your data distributions are similar or not in respect to their variances either because similar variances are required for some other analysis or because you are interested in comparing the variances of several distributions for their own sakes. It might be that the frequency distributions of heights of plants in your control group (A) in the plant growth experiment had a mean of 6 and your experimental group (B) had a mean height of 8, but the variances were 4 and 1. ($sA=2$, $sB=1$) Your distribution curves might look like these:

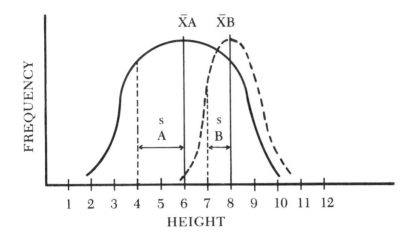

In this instance you would need a statistical test for **differences in variance** to evaluate the following null hypothesis:

$$H_0 : s^2A = s^2B \quad \text{or} \quad H_0 : 4 = 1$$

The size of the variances often has an effect on your ability to determine statistically whether a significant difference exists between two means. In the plant growth experiment just mentioned, the control had a $\bar{X}$ (mean) of 6 and the experimental group had a $\bar{X}$ of 8. Whether or not you could reject the null hypothesis that the means were the same would be strongly affected by the size of the variance *even* when they were similar (and not clearly dissimilar as above). The next figure illustrates this point. In the situation on the left, groups A and B have the same variance and means, 6 and 8. On the right, the means are again 6 and 8, the variances are also still equal, but the variances are so much smaller that you might be able to reject your null hypothesis that $\bar{X}A = \bar{X}B$.

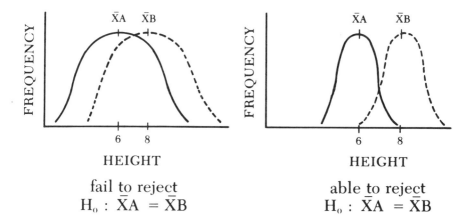

fail to reject
$H_0 : \bar{X}A = \bar{X}B$

able to reject
$H_0 : \bar{X}A = \bar{X}B$

As you study the flow chart for the selection of statistical tests, you will notice that sometimes you need to make a distinction between parametric and non-parametric statistics. In general, **non-parametric** tests can be used to evaluate null hypotheses related to any type of data distribution (normal, skewed, bimodal, etc.); they are not limited to normal distributions.

Parametric statistics are limited to normal distributions. Some statistics limit their application to instances when the following three conditions have been met (although others would relax these requirements somewhat):

1. the data must fall into a statistically normal frequency distribution;
2. the individual observations (or data points) must be independent of each other (you could not use parametric statistics when comparing weights of horse front limbs to hind limbs because the weights would be related, but you could use them to compare front limb weights of race horses to those of draft horses); and

3. the observations must be distributed on the same continuous scale of measurement.

Although it is more difficult to meet these criteria, parametric statistics should be chosen whenever possible because they are often more "powerful" tests. This means that, for a given sample size, they involve less risk that you might fail to reject a null hypothesis that really was false. (They have a greater ability to permit you to reject your null hypothesis).

Once the flow chart has been followed and the most suitable statistical treatment selected, there is one other decision that is normally made by a scientist. He must decide how much risk he is willing to take that he may come to the wrong conclusion about his null hypothesis.

When you are using a statistical test to evaluate a null hypothesis, there are two types of errors you can make; they are traditionally called the Type I and Type II errors. **A Type I error is the rejection of a true H_0. A Type II error is the failure to reject a false H_0.** For example, you might flip a *fair* coin and get ten heads in a row. Statistically this is unlikely, so your H_0: 10:0=50:50 would be rejected and you would conclude from that that the coin was unfair. This is a reasonable mistake, based on a small sample of particularly unusual data. It could happen and it would be a Type I error. You might have thrown 4 heads and 6 tails — a much more common occurrence, and failed to reject the null hypothesis that 4:6=50:50. In this instance you would conclude that the coin was fair. However, it could happen that a much larger sample size, giving results of 400:600, would have caused you to reject the null. If the coin really was unfair but you had concluded that it was fair on the basis of your toss of 4:6, you would have made a Type II error by failing to reject a false null. You can see by these examples that sample size can make a big difference. If you make a Type I error, you jeopardize your integrity as a scientist. A Type II error can also be serious, especially in claims such as one that smoking has no effect on human health.

You can never be 100% certain that you have made no mistake, but you do have some control over how certain you will be. In most statistical tests you will need, you will probably choose to set your "alpha level" at .05. **Alpha, or α, is defined as the risk of making a Type I error.** When alpha is set at .05, the chances are 1 out of 20 that you will accidentally reject a true null. These may seem like pretty favorable odds, but in some testing situations, a 5% chance of failure may be too risky. You have the option of setting any alpha level you wish *before* you calculate your test. However, a lower alpha level will increase the risk that you will make a Type II error and fail to reject a null hypothesis that really is false. The normal compromise between the risk of Type I and Type II errors is the .05 alpha level.

To illustrate this concept, let us collect data from a *fair* coin. If one test were made up of 100 tosses, and the test was repeated 10,000 times, you would get a frequency distribution of results looking more or less like the one which follows. Each test result ratio (45:65, 42:68, 30:70, 80:20, etc.) would be plotted on a frequency distribution with a center point of 50:50 and the two extremes of 100:0 and 0:100 like this:

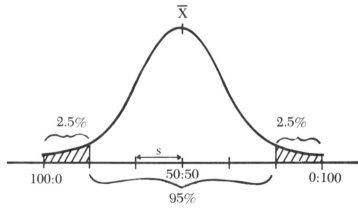

HEADS : TAILS   TAILS : HEADS

In 95% of the tests, the results will fall inside the white part of the curve (within two standard deviations from the mean); and in 5% of the tests they will fall into the shaded areas (outside two standard deviations from the mean).

Now let us suppose that you performed the 100 flip test only once, using a different coin, and that you were testing the null hypothesis that there was no meaningful difference between the results of your flipping and the theoretical ratio of 50:50. With an alpha level of .05, it *could* happen that you would flip 99 heads to 1 tail, reject your null hypothesis, and conclude that the coin was unfair when it actually was a fair coin. (That is the Type I error.) However, the odds are greater than 95 to 5 that such an extreme reading would *not* be the result of chance alone and that the coin really was unfair. This is the risk you knowingly take when you use a statistical test to evaluate whether or not you can reject your null hypothesis.

The best way to avoid making a mistake, particularly a Type II error of failing to reject a null that really is false, is to have an adequate sample size. All of the statistical tests are highly sensitive to sample size. If you are unable to reject your null hypothesis at the α level of .05, you cannot "fudge" your results by changing the α to .09; but you always have the recourse of collecting more data. If the null hypothesis really is false, the increase in sample size will eventually permit you to reject it.

Conclusions

Once you have completed your study and analyzed your results for their statistical significance, it will be up to you to draw your conclusions. There is nothing magical about the results of a statistical test, although you should be very hesitant to draw any conclusion about differences between sets of data when there is no statistical evidence for this conclusion. There may be times when statistically significant results may be biologically meaningless — particularly in the case of a poor experimental design. You, as a scientist with your own powers of reasoning, are the most important factor in any study.

The next chapter contains the flow chart to aid in the choice of statistical tests. See the various books on statistics in the bibliography of this handbook for further information and additional tests.

Chapter Seven
Statistical Tests

This chapter opens with the flow chart for helping you choose the proper statistical test. Chapter Six defined and explained the various steps in the selection process. Is it a question of correlations or differences? If differences, are you testing for differences between distributions, means, or variances? If you are testing for differences between distributions, you must select one of the X^2 (Chi Square) tests on the basis of the kind of experimental question you have asked. If you are testing for differences between means, you should select a test on the basis of whether you have two, or more than two, samples and whether or not you are in a position to use parametric statistics. All of the tests mentioned on the chart are illustrated in this chapter.

The chapter contains the following statistical tests:

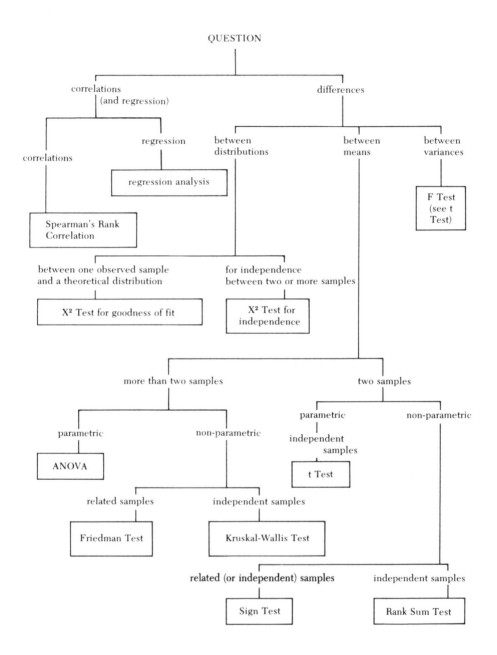

ANOVA — Analysis of Variance

Purpose:

The Analysis of Variance is a parametric statistical test of differences between means of more than two samples. It is designed to test whether the means of the various test samples vary further from the population mean than would be expected when compared to the fluctuation among observations within experimental samples.

Warnings and precautions:

1. The individual observations must be independent of each other.
2. The observations must be from a continuous scale of measurement.
3. The observations must be normally distributed.
4. The variances of the samples must be homogeneous. You may use a "quick and dirty" check for homogeneity by calculating the s^2 (variance — some calculators are programmed to do this step for you) for each sample. Then take the largest s^2 and divide it by the smallest. Proceed with the F test.

Null hypothesis:

The hypothesis to be tested is always:

$$H_0 : \bar{X}A = \bar{X}B = \bar{X}C = \bar{X}D \ldots \text{etc.}$$

If you reject the null hypothesis, you may conclude that the means are unequal when compared to the normal variation within the samples. You have not shown *which* treatment or sample is causing this to be true, however. A least-significant-difference test (LSD), found in advanced texts, will allow you to determine where the significant differences are.

Notation:

The test statistic to be computed is an F ratio. This ratio can be verbally expressed this way:

F = the average of the squared differences between the means of the samples and the grand mean (or mean of all the observations) divided by the variance of all the observations.

k = the number of test situations or treatments
n = the number of observations or scores in each treatment
N = total number of observations or scores
x = a data point or score
x^2 = x times x
SS_b = the sum of the squares between groups or treatments (don't worry about what this means)
V_b = the variance between groups or treatments
SS_w = the sum of the squares within groups or treatments (again, don't worry about what it means)
V_w = the variance within groups or treatments
SS_T = total sum of the squares
Σ = the sum

The test statistic, or F ratio we explained verbally, looks like this:

$$F = \frac{V_b}{V_w}$$

To find the values for the numerator and denominator of this formula, we first calculate SS_T and SS_b. SS_w is then found by subtracting SS_b from SS_T. You then divide SS_b and SS_w by the appropriate degrees of freedom to get V_b and V_w.

Procedure:

Step 1. Add up x's of all the treatments to obtain Σx, or the grand total.

Step 2. Square each of the x's for all treatments and then sum these squares to obtain Σx^2. (Note: on some calculators, Steps 1 and 2 can be done at the same time.)

Step 3. Compute the sums and means of the observations in each sample or treatment. In other words, add up each of the x's for each treatment and divide by n (the number of observations in that treatment).

Step 4. Compute SS_T by the following formula:
$$SS_T = \Sigma x^2 - \frac{(\Sigma x)^2}{N}$$
where $(\Sigma x)^2$ is the square of the grand total calculated in Step 1.

Step 5. Compute SS_b by the following formula:
$$SS_b = \left[\frac{(\Sigma x_A)^2}{n_A} + \frac{(\Sigma x_B)^2}{n_B} + \frac{(\Sigma x_C)^2)}{n_C} \text{ etc.}\right] - \frac{(\Sigma x)^2}{N}$$

This means square the sum of each treatment. (A, B, C etc. stand for the different treatments.) Divide by the number of observations in that treatment, and then sum for each treatment. When this is done, subtract the figure $\frac{(\Sigma x)^2}{N}$ which you calculated in Step 4.

Step 6. Compute SS_w. #$SS_w = SS_T - SS_b$

Step 7. Divide SS_b by its degrees of freedom which is equal to $k - 1$ or one less than the number of treatments.
$$V_b = \frac{SS_b}{k - 1}$$

Step 8. Divide SS_w by its degrees of freedom, which is equal to $N - k$ or the total number of observations less the number of treatments.
$$V_w = \frac{SS_w}{N - k}$$

You now have calculated the V_b and V_w values, which are the numerator and the denominator of the F ratio. Calculate F and turn to the F table on p. 57. To compare your calculated F with the table, you must consider the degrees of freedom in both the numerator and denominator of the F ratio. The degrees of freedom for the numerator (V_b) is $k - 1$, or one less than the number of treatments. The degrees of freedom for the denominator (V_w) is $N - k$, or the total number of observations less the number of treatments. You will note from the table that the degrees of freedom for the "greater variance" (usually the numerator) is read across the top. The degrees of freedom for the lesser variance, usually the denominator, will be found in the left-hand column of the table. It may be necessary to interpolate between rows to obtain an exact value. If the F statistic you have calculated is *greater* than the F value in the table (or equal to it), you have a statistically significant reason to reject your null hypothesis.

Example:

In the table below we have four sets of data (five observations in each of four test situations) — don't worry about what the experiment was. The question here, as always, is $H_0 : \bar{X}A = \bar{X}B = \bar{X}C = \bar{X}D$. We will work out the steps of the ANOVA test to see if there is enough difference between the means of these samples to permit us to reject the null hypothesis.

Raw data:

Sample A	Sample B	Sample C	Sample D
114 grams	119 grams	112 grams	117 grams
115 ”	120 ”	116 ”	117 ”
111 ”	119 ”	116 ”	114 ”
110 ”	116 ”	115 ”	112 ”
112 ”	116 ”	112 ”	117 ”

Step 1. $\Sigma x = 114 + 115 + \ldots 112 + 117 = 2{,}300$.

Step 2. $\Sigma x^2 = (114)^2 + (115)^2 + \ldots$
$(112)^2 + (117)^2 = 264{,}652$

Step 3.

Sample A	Sample B	Sample C	Sample D
$\Sigma x_A = 562$	$\Sigma x_B = 590$	$\Sigma x_C = 571$	$\Sigma x_D = 577$
$\bar{X}_A = 112.4$	$\bar{X}_B = 118.0$	$\bar{X}_C = 114.2$	$\bar{X}_D = 115.4$
$n_A = 5$	$n_B = 5$	$n_C = 5$	$n_D = 5$

Step 4. $SS_T = \Sigma x^2 - \dfrac{(\Sigma x)^2}{N}$

$N = 20$ (5 observations in each of the 4 treatments)

$SS_T = 264{,}652 - \dfrac{(2{,}300)^2}{20} = 264{,}652 - 264{,}500 = 152$

Step 5.

$$SS_b = \left[\dfrac{(\Sigma x_A)^2}{n_A} + \dfrac{(\Sigma x_B)^2}{n_B} + \dfrac{(\Sigma x_C)^2}{n_C} + \dfrac{(\Sigma x_D)^2}{n_D}\right] - \dfrac{(\Sigma x)^2}{N}$$

$$SS_b = \left[\dfrac{(562)^2}{5} + \dfrac{(590)^2}{5} + \dfrac{(571)^2}{5} + \dfrac{(577)^2}{5}\right] - 264{,}500$$

$SS_b = (63{,}168.8 + 69{,}620 + 65{,}208.2 + 66{,}585.8) - 264{,}500$

$SS_b = 264{,}582.8 - 264{,}500 = 82.8$

Step 6. $SS_w = SS_T - SS_b \quad SS_w = 152 - 82.8 = 69.2$

Step 7.

$V_b = \dfrac{SS_b}{k-1}$ (there are 4 treatments so $k - 1 = 3$)

$V_b = \dfrac{82.8}{3} = 27.60$

Step 8.

$V_w = \dfrac{SS_w}{N-k}$ ($N = 20$, $k = 4$, so $N - k = 16$)

$V_w = \dfrac{69.2}{16} = 4.325$

Now that we have V_b and V_w,

$$F = \dfrac{27.6}{4.325} \quad \text{or} \quad F = 6.38.$$

Now that we have calculated the test statistic F, we need to consult the table of F values for the two different degrees of freedom (three from the numerator and 16 from the denominator). Reading across the top of the table to 3, and down the side to 16, we find that the F value at the .05 alpha level of confidence is 3.24. Since our calculated F is much greater, we have ample reason to reject our null hypothesis.

Try this example:

Suppose an experimenter was interested in the importance of three sites in the brain controlling aggressive behavior. In his experiment he tested 4 groups of rats. In three of the groups, he lesioned one of 3 different sites in the brain. The fourth group was a control receiving no lesions. Then, working within groups receiving similar treatment, he caged his rats in pairs and observed the number of aggressive encounters that occurred in two hours. Let us suppose that these are his data. (Data are taken from Terrace and Parker; see bibliography.)

Number of fights in two hours.

Group A lesion site 1	Group B lesion site 2	Group C lesion site 3	Group D control-no lesion
12	12	19	9
14	16	18	5
8	15	25	6
11	13	21	8
15	19	22	3
12			5
10			
14			

Table of critical values for the statistic F calculated in the ANOVA test.[1]
The alpha level is .05.

Degrees of freedom for the greater variance (usually the numerator)

	1	2	3	4	5	6	7	8	9	10	12	15	20	24	30	40	60	120	∞
1	161.4	199.5	215.7	224.6	230.2	234.0	236.8	238.9	240.5	241.9	243.9	245.9	248.0	249.1	250.1	251.1	252.2	253.3	254.3
2	18.51	19.00	19.16	19.25	19.30	19.33	19.35	19.37	19.38	19.40	19.41	19.43	19.45	19.45	19.46	19.47	19.48	19.49	19.50
3	10.13	9.55	9.28	9.12	9.01	8.94	8.89	8.85	8.81	8.79	8.74	8.70	8.66	8.64	8.62	8.59	8.57	8.55	8.53
4	7.71	6.94	6.59	6.39	6.26	6.16	6.09	6.04	6.00	5.96	5.91	5.86	5.80	5.77	5.75	5.72	5.69	5.66	5.63
5	6.61	5.79	5.41	5.19	5.05	4.95	4.88	4.82	4.77	4.74	4.68	4.62	4.56	4.53	4.50	4.46	4.43	4.40	4.36
6	5.99	5.14	4.76	4.53	4.39	4.28	4.21	4.15	4.10	4.06	4.00	3.94	3.87	3.84	3.81	3.77	3.74	3.70	3.67
7	5.59	4.74	4.35	4.12	3.97	3.87	3.79	3.73	3.68	3.64	3.57	3.51	3.44	3.41	3.38	3.34	3.30	3.27	3.23
8	5.32	4.46	4.07	3.84	3.69	3.58	3.50	3.44	3.39	3.35	3.28	3.22	3.15	3.12	3.08	3.04	3.01	2.97	2.93
9	5.12	4.26	3.86	3.63	3.48	3.37	3.29	3.23	3.18	3.14	3.07	3.01	2.94	2.90	2.86	2.83	2.79	2.75	2.71
10	4.96	4.10	3.71	3.48	3.33	3.22	3.14	3.07	3.02	2.98	2.91	2.85	2.77	2.74	2.70	2.66	2.62	2.58	2.54
11	4.84	3.98	3.59	3.36	3.20	3.09	3.01	2.95	2.90	2.85	2.79	2.72	2.65	2.61	2.57	2.53	2.49	2.45	2.40
12	4.75	3.89	3.49	3.26	3.11	3.00	2.91	2.85	2.80	2.75	2.69	2.62	2.54	2.51	2.47	2.43	2.38	2.34	2.30
13	4.67	3.81	3.41	3.18	3.03	2.92	2.83	2.77	2.71	2.67	2.60	2.53	2.46	2.42	2.38	2.34	2.30	2.25	2.21
14	4.60	3.74	3.34	3.11	2.96	2.85	2.76	2.70	2.65	2.60	2.53	2.46	2.39	2.35	2.31	2.27	2.22	2.18	2.13
15	4.54	3.68	3.29	3.06	2.90	2.79	2.71	2.64	2.59	2.54	2.48	2.40	2.33	2.29	2.25	2.20	2.16	2.11	2.07
16	4.49	3.63	3.24	3.01	2.85	2.74	2.66	2.59	2.54	2.49	2.42	2.35	2.28	2.24	2.19	2.15	2.11	2.06	2.01
17	4.45	3.59	3.20	2.96	2.81	2.70	2.61	2.55	2.49	2.45	2.38	2.31	2.23	2.19	2.15	2.10	2.06	2.01	1.96
18	4.41	3.55	3.16	2.93	2.77	2.66	2.58	2.51	2.46	2.41	2.34	2.27	2.19	2.15	2.11	2.06	2.02	1.97	1.92
19	4.38	3.52	3.13	2.90	2.74	2.63	2.54	2.48	2.42	2.38	2.31	2.23	2.16	2.11	2.07	2.03	1.98	1.93	1.88
20	4.35	3.49	3.10	2.87	2.71	2.60	2.51	2.45	2.39	2.35	2.28	2.20	2.12	2.08	2.04	1.99	1.95	1.90	1.84
21	4.32	3.47	3.07	2.84	2.68	2.57	2.49	2.42	2.37	2.32	2.25	2.18	2.10	2.05	2.01	1.96	1.92	1.87	1.81
22	4.30	3.44	3.05	2.82	2.66	2.55	2.46	2.40	2.34	2.30	2.23	2.15	2.07	2.03	1.98	1.94	1.89	1.84	1.78
23	4.28	3.42	3.03	2.80	2.64	2.53	2.44	2.37	2.32	2.27	2.20	2.13	2.05	2.01	1.96	1.91	1.86	1.81	1.76
24	4.26	3.40	3.01	2.78	2.62	2.51	2.42	2.36	2.30	2.25	2.18	2.11	2.03	1.98	1.94	1.89	1.84	1.79	1.73
25	4.24	3.39	2.99	2.76	2.60	2.49	2.40	2.34	2.28	2.24	2.16	2.09	2.01	1.96	1.92	1.87	1.82	1.77	1.71
26	4.23	3.37	2.98	2.74	2.59	2.47	2.39	2.32	2.27	2.22	2.15	2.07	1.99	1.95	1.90	1.85	1.80	1.75	1.69
27	4.21	3.35	2.96	2.73	2.57	2.46	2.37	2.31	2.25	2.20	2.13	2.06	1.97	1.93	1.88	1.84	1.79	1.73	1.67
28	4.20	3.34	2.95	2.71	2.56	2.45	2.36	2.29	2.24	2.19	2.12	2.04	1.96	1.91	1.87	1.82	1.77	1.71	1.65
29	4.18	3.33	2.93	2.70	2.55	2.43	2.35	2.28	2.22	2.18	2.10	2.03	1.94	1.90	1.85	1.81	1.75	1.70	1.64
30	4.17	3.32	2.92	2.69	2.53	2.42	2.33	2.27	2.21	2.16	2.09	2.01	1.93	1.89	1.84	1.79	1.74	1.68	1.62
40	4.08	3.23	2.84	2.61	2.45	2.34	2.25	2.18	2.12	2.08	2.00	1.92	1.84	1.79	1.74	1.69	1.64	1.58	1.51
60	4.00	3.15	2.76	2.53	2.37	2.25	2.17	2.10	2.04	1.99	1.92	1.84	1.75	1.70	1.65	1.59	1.53	1.47	1.39
120	3.92	3.07	2.68	2.45	2.29	2.17	2.09	2.02	1.96	1.91	1.83	1.75	1.66	1.61	1.55	1.50	1.43	1.35	1.25
∞	3.84	3.00	2.60	2.37	2.21	2.10	2.01	1.94	1.88	1.83	1.75	1.67	1.57	1.52	1.46	1.39	1.32	1.22	1.00

Degrees of freedom for the lesser variance (usually the denominator)

1. Adapted from R.G.D. Steel and J.H. Torrie. Principles and Procedures of Statistics. McGraw-Hill Book Company, New York. 1960.

The Friedman Test

Purpose:

The Friedman Test, a non-parametric test of differences between means, is used to test for significant differences between the responses of several matched samples (paired samples) exposed to three or more treatments. This experimental design is often called a randomized block design. The Friedman Test is the non-parametric equivalent of the ANOVA test.

Warnings and precautions:

1. Data must be measured on an ordinal, interval, or ratio scale.
2. The items in each sample must be of a matched design. Hence the same number of items will be in each group and must be randomly assigned to each treatment (see the example).

Procedure:

The null hypothesis to be tested is that the means of the responses to three or more treatments do not differ significantly from each other.
(H_o: $\bar{X}A = \bar{X}B = \bar{X}C$ . . . etc.)

1. Place the response values (or scores) in a two-way table that has k columns (k = the number of samples or treatments) and b rows or blocks (b = the number of observations or samples in each treatment).
2. Rank the values in each row from 1 to however many columns you have. If this is not clear, see the example and also the discussion of ranking data in the Spearman's Rank Correlation Test. Tied values each receive the average rank of the ranks that would have been assigned to those tied values.
3. Add up the ranks found in each column producing R_A (the sum of the ranks in column or treatment A), R_B, R_C, . . . etc.

4. Calculate the degrees of freedom according to the formula, $df = k - 1$ (k = the number of treatments or columns).
5. Calculate the value of the test statistic according to this formula:

$$X_r^2 = \frac{12}{bk(k+1)} \Sigma(R_s)^2 - 3b(k+1)$$

b = the number of rows or observations
k = the number of columns or treatments

$\Sigma(R_s)^2$ = Take the sum of each column's ranks and square it. Add all the squared totals of the ranks of each column.

6. For small values of b and k, use the accompanying table to check your calculated test statistic against the critical values. For values not found on this table, use the X^2 table found on page 95.
7. If the calculated value of X_r^2 is equal to or greater than the critical value in the table, you may reject your null hypothesis and accept the alternative hypothesis, which is that the means of the responses to the treatments are *not* equal. If the calculated test statistic is less than the critical value on the table you are using, you fail to reject your null hypothesis.

Example:

An experimenter wishes to study potential stress in mice (as measured by weight loss) caused by smoke inhalation, bright light exposure, and exposure to rock music. He takes 9 cages, each containing 3 mice which are littermates of the same age and sex, and randomly assigns one of the three mice from each cage to each of the following treatment groups:

Treatment A — mice which are placed in a chamber full of cigar smoke for one hour each day.

Treatment B — mice which are placed in very bright light for 1 hour each day.

Treatment C — mice which are exposed to rock music for 1 hour each day.

Except for the time when the mice are exposed to the test conditions, the mice are maintained in the same fashion.

The following table shows the weight loss in grams for each of the mice exposed to each treatment.

Cage #	Treatment A (smoke)		Treatment B (light)		Treatment C (music)	
	wt. loss	rank	wt. loss	rank	wt. loss	rank
1	5	(2)	4	(1)	6	(3)
2	2.5	(3)	1	(1)	2	(2)
3	4	(1)	5	(2)	7	(3)
4	2	(2)	1	(1)	3	(3)
5	4	(2)	3	(1)	5.5	(3)
6	1	(1.5)	1	(1.5)	2	(3)
7	4	(2)	3.5	(1)	5	(3)
8	4	(1)	5	(3)	4.5	(2)
9	1.5	(1)	2	(2)	3	(3)
sum of the ranks		15.5		13.5		25

$H_0 : \bar{X}A = \bar{X}B = \bar{X}C$

calculations:

$$X_r^2 = \frac{12}{bk(k+1)} \; \Sigma \; (R_s)^2 \; - \; 3b(k+1)$$

$$X_r^2 = \frac{12}{(9)\,(3)\,(3+1)}[\,(15.5)^2 + (13.5)^2 + (25)^2\,] - (3)\,(9)\,(3+1)$$

$X_r^2 = 8.39$

Using the table on page 62, the critical value for this example, with an b of 9 and a k of 3, is 6.222. Since our calculated test statistic X_r^2 is 8.39, which is greater than the critical value of 6.222, we are able to reject our null hypothesis in favor of the alternative hypothesis. (Note that you have only shown that the means are unequal — not which treatment causes this to be so.)

Try this example:

In a test comparing three commercial fertilizers, 4 kernels of corn were randomly selected from each of 12 ears of Illini Chief corn. One of each set of four was subjected to each of the following experimental situations. Group 1 was planted in soil with no added fertilizer, group 2 was planted in similar soil with Fertilizer A, group three with Fertilizer B, and group 4 with Fertilizer C. All kernels were planted on the same day and subjected to identical environmental conditions. After 2 weeks, the growth in centimeters was measured for each young plant. The data looked like this:

Growth in Centimeters

Ear number	No fertilizer	Fertilizer "A"	Fertilizer "B"	Fertilizer "C"
1	10cm	12cm	13 cm	11cm
2	11.5	11	14	10
3	9	16	17	15
4	12	12	14	16
5	15	18	17	16
6	16	20	22	24
7	11	21	23	24
8	10.5	14	18	16
9	9	20	22	10
10	14	15	11	17
11	13	25	20	18
12	12	14	15	16

Is there a statistical reason to reject the null hypothesis that the mean growth under the four experimental situations is the same? (Note that with this larger sample size you will have to use the X^2 table found on page 95.)

Table of critical values for the Friedman Test at the .05 alpha level.[1]

k	b	Critical Value
3	3	6.000
3	4	6.500
3	5	6.400
3	6	7.000
3	7	7.143
3	8	6.250
3	9	6.222
4	2	6.000
4	3	7.400
4	4	7.800

(for larger sample sizes, use the X^2 table on page 95, with K-1 d.f.)

1. M. Friedman, 1937. The use of ranks to avoid the assumption of normality implicit in the analysis of variance. *J. Am. Stat. Assoc.* 32:675-701.

The Kruskal-Wallis Test

Purpose:

The Kruskal-Wallis Test is a non-parametric test for differences between means that can be used in any situation appropriate for single factor analysis of variance. However, it does not have to meet all the requirements of a parametric test. It is appropriate for comparing more than two independent samples of equal or unequal size.

Warnings and precautions:

1. The measurement scale must be an ordinal, interval, or ratio scale.
2. All random variables should be continuous.
3. A large number of tied values may distort the true level of significance.
4. The more-than-two samples must be independent of each other and the values within each sample must also be independent.

Notation:

k = the number of samples (the number of groups compared)

n_1 = the number of items in the first sample

n_2 = the number of items in the second sample

n_3 = the number of items in the third sample, etc.

N = the total number of observations in all samples ($n_1 + n_2 + n_3 + n_4 + n_5$ etc.)

R_1 = the sum of the ranks assigned to the values in the first sample

R_2 = the sum of the ranks assigned to the values in the second sample, etc.

H = is the test statistic you are going to calculate

The null hypothesis:

The means of the k samples do not differ significantly from each other. ($\bar{X}_1 = \bar{X}_2 = \bar{X}_3 = \bar{X}_4$ etc.)

63

The alternative hypothesis is that at least one of the samples contains values that are significantly different from the others. (Note that with this test you conclude *only* that the samples are unequal — you do not determine which one or more is different.)

Procedure:

1. Rank all of the N observations into a single series by assigning the rank of 1 to the smallest value, etc. Rank without regard to group. Ties are assigned the average value of the ranks that would have been assigned to them. (See the example and also the discussion of ranking data in the Spearman's Rank Correlation test.)
2. Calculate R for each of the k samples.
3. Calculate the Kruskal-Wallis test statistic H by the following formula:

$$H = \frac{12}{N(N+1)} \left[\frac{R_1{}^2}{n_1} + \frac{R_2{}^2}{n_2} + \frac{R_3{}^2}{n_3} \text{ etc.} \right] - 3(N+1)$$

4. If k = 3, *and* all sample sizes are 5 or less, use the Kruskal-Wallis test table for comparing your calculated H to the critical value.

   If k = 4 or more, *or* the sample sizes are greater than 5, use the X^2 Table with k-1 degrees of freedom (p. 95.)

If the test statistic H is equal to or greater than the critical value in either table, you may reject your null hypothesis.

Example:

Four farmers each claim to have developed a better way of raising corn. They decide to use the number of ounces of dried corn kernels per plant as a measure of yield. Each farmer picks one good plant from each row for his measurement. The farms differ in number of rows. The table of data on the next page presents the ounces per plant and the rank for each value.

FARM A		FARM B		FARM C		FARM D	
weights	rank	weights	rank	weights	rank	weights	rank
28	4	63	35	63	35	63	35
32	7	55	28.5	48	23	69	40
25	1	66	38	26	2	56	30
36	12	58	31	33	8.5	61	33
36	12	67	39	41	17.5	45	21
36	12	55	28.5	37	14	52	26
40	16	47	22	29	5	59	32
33	8.5	51	25	34	10	53	27
50	24	42	19	30	6	65	37
44	20			38	15		
				27	3		
				41	17.5		

$R_1 = 116.5$ $R_2 = 266$ $R_3 = 156.5$ $R_4 = 281$
$n_1 = 10$ $n_2 = 9$ $n_3 = 12$ $n_4 = 9$

$N = 40$
$k = 4$
degrees of freedom $= 3$
This is the calculation of the test statistic H:

$$H = \frac{12}{40(40 + 1)}\left[\frac{116.5^2}{10} + \frac{266^2}{9} + \frac{156.5^2}{12} + \frac{281^2}{9}\right] - 3(40 + 1)$$

or:

$$H = \frac{12}{1640}\left[\frac{13572.25}{10} + \frac{70756}{9} + \frac{24492.25}{12} + \frac{78961}{9}\right] - 123$$

or:

$H = .007317(1357.225 + 7861.78 + 2041.02 + 8773.44) - 123$
$H = .007317 (20033.465) - 123$ and finally,
$H = 23.59$

When the H of 23.59 is compared to the critical value on the X^2 table at 3 degrees of freedom, we find that 23.59 is larger than 7.81 and therefore we may reject our null hypothesis. In this case, we may conclude only that one or more of the farmer's samples of corn are significantly different from the others. We can gain some idea of how the samples differ by calculating an average weight (note that the sample sizes are unequal) but this statistical test does not allow us to draw any conclusions about how they differ. Other statistical tests exists which permit you to discriminate more finely.

(In our example we had more than three samples and more than five values in each sample so we compared our test statistic H to critical values in the X^2 table. If there had been only 3 samples and the sample sizes had been 5 or less, we would have used the table of critical values given on page 67.)

Try this example:

Three first grade teachers, who use different teaching techniques, decided to compare their methods for teaching reading. Each selected a test group of students who could not read at the beginning of the school year, taught them in their usual way, and then gave them a standard reading test at the end of the year. (They assumed that all the test students had an equal learning ability.)

Teacher A's students had the following test scores:
60 75 72 63 50 80 88 90 (n_1 = 8, average score = 72.25)

Teacher B's students had the following scores:
55 65 82 86 90 92 60 58 95 70 (n_2 = 10, average score = 75.3)

Teacher C's students had these scores:
48 52 92 85 60 62 65 75 80 58 76 78 (n_3 = 12, average = 69.25)

Is there any statistical reason, based on these test scores, to believe that one teaching method is any better than another? In other words, the null hypothesis is that there is no measurable difference in performance between the students taught in the different methods.
H_0 : $\bar{X}A$ = $\bar{X}B$ = $\bar{X}C$

Table of critical values at the .05 level for the Kruskal-Wallis Test[1]

Sample Sizes

n_1	n_2	n_3	Critical Values
3	2	2	4.714
3	3	1	5.143
3	3	2	5.361
3	3	3	5.600
4	2	1	4.8214*
4	2	2	5.333
4	3	1	5.208
4	3	2	5.444
4	3	3	5.727
4	4	1	4.967
4	4	2	5.455
4	4	3	5.598
4	4	4	5.692
5	2	1	5.000
5	2	2	5.160
5	3	1	4.960
5	3	2	5.251
5	3	3	5.648
5	4	1	4.986
5	4	2	5.273
5	4	3	5.656
5	4	4	5.657
5	5	1	5.127
5	5	2	5.338
5	5	3	5.705
5	5	4	5.666
5	5	5	5.780

(*this is actually at .057, no alpha level of .05 was available at this sample size)

1. Adapted from J.H. Zar. *Biostatistical Analysis*. Prentice-Hall, Inc. Englewood Cliffs, N.J. 1974.

The Rank Sum Test

(based on the White modification of the Wilcoxon Rank Sum Test)

Purpose:

The Rank Sum Test, a non-parametric test of differences between means, is used to test for significant differences between two samples of equal or unequal size.

Warnings and precautions:

1. This test is not appropriate for paired data.
2. Data must be measured on an ordinal, interval, or ratio scale.

Procedure:

Null hypothesis: The mean of sample A is not significantly different from that of sample B ($H_o : \bar{X}A = \bar{X}B$).

1. Rank all data points from both samples into a single series. Tied absolute values each get the average of the ranks that they would have been assigned if the values had not been tied.
2. If the sample sizes are equal, obtain the sum of the ranks for each sample. Call the smaller sum "T" and go to step 4. If the sample sizes are not equal, go to step 3.
3. Call the smaller sample size "n_1" and the larger "n_2." Obtain the rank sum of the smaller sample (n_1) by adding the ranks of its points. Call this sum T_1. Calculate T_2 by this formula:

$T_2 = n_1 (n_1 + n_2 + 1) - T_1$ where n_1 = the sample size of the smaller sample, and n_2 = the sample size of the larger sample. Let T = the smaller of T_1 or T_2.

4. For sample sizes not included on the table on page 71, go to step 5. For sample sizes included in the table, compare T with the critical value in the table. If the smaller rank sum is equal to, or *smaller*

68

than, the critical value, you may reject the null hypothesis. If the value is greater than the critical value, you fail to reject the H_0.

5. Compute the statistic Z by the following formula:

$$Z = \frac{(|m - T| - \frac{1}{2})}{S}$$

$$m = \frac{n_1 (n_1 + n_2 + 1)}{2} \quad (n_1 \text{ and } n_2 \text{ are as above})$$

$$S = \sqrt{\frac{n_2 m}{6}}$$

T = the smaller of the two rank sums (T_1 or T_2)

If Z is *equal to* or greater than 1.96, you may reject your null hypothesis. If not, you fail to reject the H_0.

Example:

A paperboy covers both a morning route to the south of town and an evening route to the north. The sizes of the dogs that attack the paperboy in the morning route are listed in Sample A and the sizes of those attacking in the evening are in Sample B. Is the paperboy attacked by larger dogs on one route than on the other? The null hypothesis would be that the average size of the dogs attacking on the southern route was equal to the average size of the dogs attacking on the northern route. ($H_0 : \bar{X}A = \bar{X}B$). In the following table we have shown how you would list your size measurements (in the order you took them) for each sample and then assign ranks to the measurements of both groups treated as one. (Rank 1 is assigned to the smallest measurement, etc.)

Sample A	Rank	Sample B	Rank
12"	3	33"	15
14"	4	21"	7
18"	5.5	23"	8.5
25"	10	23"	8.5
28"	12	46"	19
10"	2	35"	16
7"	1	31"	14
27"	11	37"	17
30"	13	42"	18
		18"	5.5
sum of ranks	61.5	sum of ranks	128.5

Sample A has only 9 readings; it is the smaller sample size, or n_1. T_1 is the sum of the ranks of the smaller sample (61.5). T_2 = the first sample size, n_1 (or 9), times the sum of the two sample sizes (19) plus 1, or 9 x 20, less T_1. 9 x 20 = 180. 180 − 61.5 = 118.5. When T_1 (which is 61.5) is compared with T_2 (118.5), we find that T_1 is the *smaller* and therefore is the figure T which we must compare with the critical value for our two sample sizes on the chart on page 71. Reading across the top line to our n_1 or 9, and down the side to our n_2 which is 10, we find the critical value is 65. Our T (61.5) is smaller than the critical value 65; therefore we may reject our null hypothesis that the samples were equal.

Try this example:

In order to examine the potential ability of vitamin C to prevent colds, a experimental group of 15 teenagers (selected at random from a group of 30) was given 10 g of vitamin C per day for one year. A control group of the remaining 15 were given similar-looking, harmless pills which contained no vitamin supplement. In other respects, each youth maintained his normal diet. Each recorded the number of colds he had during the test year.
These are the results from the experimental group receiving vitamin C (n_1 = 15):
    4 2 3 1 5 4 0 2 1 6 0 2 3 1 2 (an average of 2.4 colds/yr.)
These are the numbers of colds recorded by the control group (n_2 = 15):
    4 5 2 6 8 7 10 9 5 4 6 1 1 3 4 (average number of colds = 5)
Is there a statistically significant reason to conclude that vitamin C is effective in cold prevention? $H_0 : \bar{X}_{exp.} = \bar{X}_{contr.}$

Table of critical values for the Rank Sum Test at the .05 alpha level.[1]

sample sizes for n_1

n_2	2	3	4	5	6	7	8	9	10	11	12	13	14	15
4			10											
5		6	11	17										
6		7	12	18	26									
7		7	13	20	27	36								
8	3	8	14	21	29	38	49							
9	3	8	15	22	31	40	51	63						
10	3	9	15	23	32	42	53	65	78					
11	4	9	16	24	34	44	55	68	81	96				
12	4	10	17	26	35	46	58	71	85	99	115			
13	4	10	18	27	37	48	60	73	88	103	119	137		
14	4	11	19	28	38	50	63	76	91	106	123	141	160	
15	4	11	20	29	40	52	65	79	94	110	127	145	164	185
16	4	12	21	31	42	54	67	82	97	114	131	150	169	
17	5	12	21	32	43	56	70	84	100	117	135	154		
18	5	13	22	33	45	58	72	87	103	121	139			
19	5	13	23	34	46	60	74	90	107	124				
20	5	14	24	35	48	62	77	93	110					
21	6	14	25	37	50	64	79	95						
22	6	15	26	38	51	66	82							
23	6	15	27	39	53	68								
24	6	16	28	40	55									
25	6	16	28	42										
26	7	17	29											
27	7	17												
28	7													

sample sizes for n_2

1. Adapted from R.G.D. Steel and J.H. Torrie. *Principles and Procedures of Statistics*. McGraw-Hill Book Co., N.Y. 1960.

Regression Analysis (Simple Linear Regression)[1]

Purpose:

The Regression Analysis tests to see whether or not there is a functional relationship between variables. In simple linear regression, we are dealing with only two variables and testing to see whether the functional relationship between the dependent variable Y and the independent variable X can be described as a straight line. This type of analysis can be used to examine causal relationships between variables and to predict one variable given the value of the other.

The hypothesis to be tested is whether the slope of the line in the equation $Y = a + bX$ is, for statistical purposes, equal to 0.

$H_0 : b = 0.$

The regression equation $Y = a + bX$ describes a relationship in which Y is the dependent variable, X is the independent variable, and b is the slope of the line which describes the change in Y per unit change in X. And "a" is the point where the line crosses the Y axis, or the "Y intercept." This relationship is illustrated in this figure:

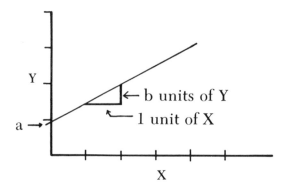

1. The authors thank Nicholas J. Cheper, Department of Zoology, University of Tennessee/Knoxville, for contributing this test.

Warnings and Precautions:

1. The test assumes that the values of X, the independent variable, are measured without error. (The values of X are fixed by the experimenter and only the values of Y are free to vary.)
2. The Y values must be measured on a continuous scale. The X values can be discrete or continuous but usually are discrete.
3. The test assumes that values of Y are from a normal distribution.
4. It also assumes that the variance around the regression line is the same for all XY pairs.

Notation:

ΣX = the sum of the values for X, or the independent variable

ΣX^2 = the sum of the squares of each independent variable

ΣY = the sum of the values for Y, the dependent variable

ΣY^2 = the sum of the squares of the dependent variables

ΣXY = multiply each X by each Y and add up the products of each XY pair

n = the number of XY pairs

SS_T = the total sum of the squares of the Y values

SS_r = the sum of the squares due to regression

SS_E = the residual sum of the squares

V_r = the variance due to regression

V_E = the variance due to residuals

Procedure:

Step 1. Arrange the data by XY pairs (see the example if this isn't clear).

Step 2. Compute ΣX by adding up all the X values, and compute the $\bar{X}$ (the mean of X) by dividing the value just calculated by the number of X readings.

Step 3. Compute ΣX^2 by squaring each X reading and adding the squares.

Step 4. Compute ΣY and $\bar{Y}$ (the mean of Y) as you did for the X values.

Step 5. Compute ΣY^2.

Step 6. Compute ΣXY.

Step 7. Compute the slope b by the following formula:

$$b = \frac{\Sigma XY - \dfrac{(\Sigma X)(\Sigma Y)}{n}}{\Sigma X^2 - \dfrac{(\Sigma X)^2}{n}}$$

Step 8. Compute a, the Y intercept where $X = 0$, by the following formula:

$$a = \bar{Y} - b\bar{X}$$

Step 9. Place the values you have calculated into this equation:

$$Y = a + bX$$

Step 10. Compute SS_T by the formula:

$$SS_T = \Sigma Y^2 - \frac{(\Sigma Y)^2}{n}$$

Step 11. Compute SS_r by the formula:

$$SS_r = b\left[\Sigma XY - \frac{(\Sigma X)(\Sigma Y)}{n}\right]$$

Step 12. Subtract SS_r from SS_T to get the SS_E.

Step 13. Now we need to divide the SS_r and the SS_E by the appropriate degrees of freedom to get the variance for each (the V_r and the V_E). The degrees of freedom for SS_r is *always* 1. Therefore, $SS_r = V_r$. The degrees of freedom for $SS_E = n - 2$. Therefore the formula for the variance due to residuals is $V_E = \dfrac{SS_E}{n - 2}$.

Step 14. The last step is to set up an F ratio using the formula

$$F = \frac{V_r}{V_E}$$

and calculate a value for F which can be compared with

the critical values for F in the table used with the ANOVA test on p. 57. The critical value for F must be found for 1 degree of freedom in the numerator and $n - 2$ degrees of freedom in the denominator. If the F value you have calculated is *equal to or greater than* the critical value in the F table, you may reject the null hypothesis that $b = 0$ and accept the alternate hypothesis that your data can be explained by a simple linear regression with a slope of b describing the functional linear relationship between your values of X and those of Y. If the $F_{calculated}$ is less than the $F_{critical}$ on the table, you fail to reject your null hypothesis and must conclude that no linear relationship exists between your two variables.

Example:

Let us suppose that you wanted to know whether the length and girth of bean shoots were related. Here are some measurements of the circumference of bean shoots (the Y values) measured for various lengths of bean shoots (the X values).

X — shoot length	Y — shoot girth
1	2.5
3	4.6
4	9.5
7	11.6
9	14.8
11	20.0

Step 1. The data are already arranged in XY pairs.

Step 2. $\Sigma X = 35 \quad \bar{X} = 5.83$

Step 3. $\Sigma X^2 = 277$

Step 4. $\Sigma Y = 63 \quad \bar{Y} = 10.5$

Step 5. $\Sigma Y^2 = 871.26$

Step 6. $\Sigma XY = 488.7$

Step 7. Inserting the figures just calculated into the formula for calculating the slope, b, we get the following equation:

$$b = \frac{488.7 - \dfrac{(35)\ (63)}{6}}{277 - \dfrac{(35)^2}{6}} \quad \text{or } b = \frac{488.7 - 367.5}{277 - 204.0}$$

$$b = \frac{121.2}{73} \quad \text{or } b = 1.664$$

Step 8. Calculate a, the Y intercept, by the formula $a = \bar{Y} - bX$.
$a = 10.5 - (1.664)\ (5.83)$ or $a = .80$

Step 9. $Y = a + bX$, so $Y = .80 + 1.664X$

Step 10. Calculate SS_T. $SS_T = 817.26 - \dfrac{(63)^2}{6}$
$SS_T = 209.76$

Step 11. Calculate SS_r.

$$SS_r = 1.664 \left[488.7 - \frac{(35)\ (63)}{6}\right] \quad \text{or } SS_r = (1.664)\ (121.2)$$

$$SS_r = 201.6768$$

Step 12. Calculate SS_E. $SS_E = 209.76 - 201.6768$
$SS_E = 8.0832$

Step 13. Calculate V_r and V_E.
$$V_r = SS_r \text{ or } 201.6768$$
$$V_E = \frac{8.0832}{4} \quad \text{or } V_E = 2.0208$$

Therefore the F ratio is

$$F = \frac{201.6768}{2.0208} \quad \text{or } F = 99.80$$

When we compare our calculated F of 99.80 to the appropriate F value on the F table (with the ANOVA test, p.57). at the appropriate degrees of freedom 1 and 4 ($n = 6$, $n - 2 = 4$), we find that our $F_{calculated}$ of 99.8 is much greater than the $F_{critical}$ of 7.71. Therefore we can reject our null hypothesis that $b = 0$ (or that there is no linear functional relationship between the height and girth of bean shoots) and conclude that a dependent relationship exists.

Try this example:

The wing lengths of sparrows were measured at various times after hatching. The data are given below, with X being the age in days after hatching and Y the wing length measured in centimeters. Is there a linear functional relationship between the values of X and Y in the following set of data (taken from J.H. Zar, see bibliography)?

Age in days X	Wing length in cm. Y
3	1.4
5	2.2
6	2.4
10	3.2
14	4.7
17	5.0

The Sign Test

Purpose:

The Sign Test, a non-parametric test of differences between means (more correctly, medians), is used to determine whether two related samples differ significantly, or whether some treatment causes a change in an experimental group.

Warnings and precautions:

1. The data must be measured on an ordinal, interval, or ratio scale.
2. The "experiment" may be of a related sample paired design.
3. The measured variable must have some sequential arrangement of magnitude.

Procedure:

Total the number of "greater than" responses. Total the number of "less than" responses. "No change" responses are ignored. Find the square on the Sign Test figure (page 80) that corresponds to these values. If this square lies in the "do not reject" region, you fail to reject your hypothesis. If it lies outside this region (in either reject region), you may reject your null hypothesis.

Example:

You measure the jumping abilities of 25 frogs. Then you give them all a dose of "Calaveras Tonic" and find that 20 of the frogs now can jump farther than they did before, and 5 jump less far. H_0 : the average jumping distance before Calaveras Tonic = the average jumping distance after the tonic for each frog. You would find the Plus 20 (20 "more than" responses)/Minus 5 square. It lies in the upper "reject" region. You may reject your hypothesis and conclude that Calaveras Tonic makes frogs jump farther.

In the Sign Test, the H_0 is: the ratio of "more than" to "less than" responses is not significantly different from 50:50.

Try this example:

The strength of the knee-jerk reflex, measured in degrees of arc, was tested for 10 men of the same age under two conditions — with the muscles tensed (T) and relaxed (R). The data (taken from J.P. Guilford; see bibliography) are presented in the table below. The null hypothesis, when testing these data with the Sign Test, is that they represent a random sample from the *same* population — or that there is no difference in the reflex response under the two test conditions. The signs (+ or −) in the table represent the changes between the tensed and relaxed conditions.

Knee-jerk reflexes, in degrees of arc, for 10 men tested under tensed (T) and relaxed (R) conditions.

T	R	sign of T − R
19	14	+
19	19	o
26	30	−
15	7	+
18	13	+
30	20	+
18	17	+
30	29	+
26	18	+
28	21	+

(Note that there are 10 pairs of observations but, in one pair, no change is involved. Therefore we have 8 pluses and 1 minus.)

Table for the Sign Test at the alpha level of .05.[1]

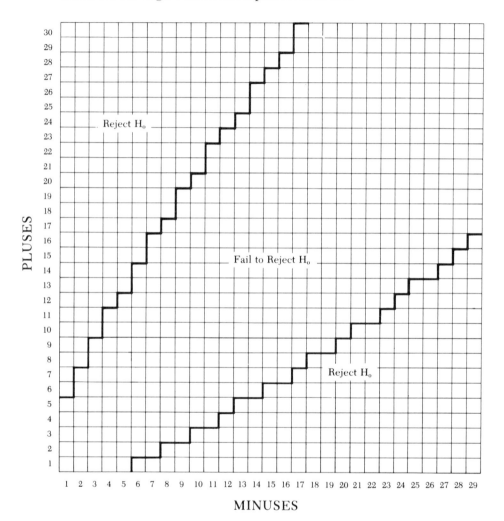

PLUSES

MINUSES

1. Constructed from values found in The Chemical Rubber Co., *Handbook of Probability and Statistics*, 2nd ed. (W.H. Beyer, ed.) The Chemical Rubber Company, Cleveland, Ohio. 1968.

Spearman's Rank Correlation

Purpose:

This non-parametric test of relationships may be used to establish whether or not two variables are correlated.

Warnings and precautions:

1. The data must be from an ordinal, interval, or ratio scale of measurement.
2. The individual data points must be independent.
3. Remember that correlation does not necessarily imply cause and effect.

Procedure:

The test statistic, is indicated by "r_s." H_0: there is no correlation between the two variables in question.

1. Separately rank the variables for each data point within the two groups. Tied absolute values each get the average rank of these two values had they not been tied. If this is unclear, look at the absolute values of some data and how they are ranked:

values:	8	12	14	16	16	20	30
ranks:	1	2	3	4.5	4.5	6	7

(here two values are tied)

or,

values:	8	12	16	16	16	20	30
ranks:	1	2	4	4	4	6	7

(three are tied)

2. Compute the differences between the ranks (d) for the two variables for each data point.
3. Square each difference.
4. Sum the square of the differences (Σd^2).
5. Apply the following formula:

$$r_s = 1 - \left(\frac{6 \, \Sigma \, d^2}{n^3 - n} \right)$$

where d^2 = the square of the differences between the ranks for the two variables that establish each point, and n = the number of individual points.

6. Compare the calculated statistic r_s with the critical value given in the table on page 84 for the appropriate sample size.

If r_s is *greater than*, or *equal to*, the critical value, you may reject the null hypothesis.

Example:

Suppose that you want to know whether there is a correlation between heights and weights of the six men listed in the following table. Your H_0 would be that there is no correlation between the height and the weight of these men. This is how you would arrange your data and rank the variables:

Individual	Height (inches)	Rank	Weight (lbs)	Rank	d	d²
Fred	68	2	140	2	0	0
Ralph	69	3	170	5	2	4
Sam	71	4	160	4	0	0
Tom	73	5	150	3	2	4
Dick	66	1	130	1	0	0
Harry	74	6	180	6	0	0
					Σd^2 =	8

This would be your calculation:

$$r_s = 1 - \left(\frac{6\Sigma d^2}{n^3 - n} \right) \qquad r_s = 1 - \left(\frac{48}{6^3 - 6} \right)$$

$$r_s = 1 - \left(\frac{48}{210} \right) \quad r_s = 1 - .23 \quad \text{therefore } r_s = .77$$

You have now calculated r_s, which is .77. Go to the rank correlation table (which we have given for the alpha level of .05 only) and find the critical value for your sample size (n) which is 6. The critical value from the table is .886. Since your calculated r_s is less than the critical value, the correlation is not significant at the .05 alpha level of confidence. You have failed to reject your null hypothesis. You cannot assume that there is a relationship between the height and the weight of these men.

Try this example:

Fifteen individuals were shown sets of limericks and sets of cartoons and asked to judge the humor value of each on a 5 point scale. In the data given in the following table, the cartoon score is the sum of the points assigned to each item in the set by one individual. The limerick score also represents that same individual's total scores for the limericks in the test set. Because there are several tied scores, we have supplied the ranks of each set for you. (The data are taken from J.P. Guilford; see bibliography.)

Cartoon score	Limerick score	R_1	R_2	d	d^2
47	75	11	8	3	9
71	79	4	6	2	4
52	85	9	5		
48	50	10	14	etc.	
35	49	14.5	15		
35	59	14.5	12		
41	75	12.5	8		
82	91	1	3		
72	102	3	1		
56	87	7	4		
59	70	6	10		
73	92	2	2		
60	54	5	13		
55	75	8	8		
41	68	12.5	11		

In this instance the Spearman's Rank Correlation Test would determine whether or not there was a statistically significant correlation between the perceived humor content in the cartoons and the limericks. The null hypothesis would be that there is no correlation between the two sets of data.

Table of critical values for different sample sizes at the .05 alpha level to be used with the Spearman's Rank Correlation test.[1]   (n = sample size)

n	critical value	n	critical value	n	critical value	n	critical value
5	1.00	27	0.382	49	0.282	92	0.205
6	0.886	28	0.375	50	0.279	94	0.203
7	0.786	29	0.368	52	0.274	96	0.201
8	0.738	30	0.362	54	0.268	98	0.199
9	0.700	31	0.356	56	0.264	100	0.197
10	0.648	32	0.350	58	0.259		
11	0.618	33	0.345	60	0.255		
12	0.587	34	0.340	62	0.250		
13	0.560	35	0.335	64	0.246		
14	0.538	36	0.330	66	0.243		
15	0.521	37	0.325	68	0.239		
16	0.503	38	0.321	70	0.235		
17	0.485	39	0.317	72	0.232		
18	0.472	40	0.313	74	0.229		
19	0.460	41	0.309	76	0.226		
20	0.447	42	0.305	78	0.221		
21	0.435	43	0.301	80	0.220		
22	0.425	44	0.298	82	0.217		
23	0.415	45	0.294	84	0.215		
24	0.406	46	0.291	86	0.212		
25	0.398	47	0.288	88	0.210		
26	0.390	48	0.285	90	0.207		

1. Adapted from J.H. Zar. *Biostatistical Analysis.* Prentice-Hall, Englewood Cliffs, N.J. 1974.

The t Test and the F Test[1]

Purpose:

The t Test is a parametric test for differences between means of independent samples. The formula for calculating t takes on different forms depending upon whether or not the two samples being compared have (statistically) equal variances. Step 1 under the procedures is the **F test** which **can establish whether or not the variances of the samples are significantly different.**

Warnings and precautions:

This test requires that:
1. The individual observations be independent of each other;
2. The distribution of observations be continuous; and
3. The observations be normally distributed.
4. If the variances of the samples differ significantly from each other, you calculate t by the formula in step 4.

Notation:

n_1 and n_2 = the sample sizes of sample 1 and sample 2, respectively.

$\bar{X}_1$ and $\bar{X}_2$ = the means of samples 1 and 2, respectively.

Procedure:

1. Determine whether or not the variances of the two samples in question are significantly different from each other by using the F test described below. The $H_0 : s^2_1 = s^2_2$ the $H_1 : s^2_1 \neq s^2_2$

To use the **F test,** you need to calculate a value F and your degrees of freedom. F is calculated by dividing the larger of the two variances by the smaller. If s^2_1 is numerically larger than s^2_2, then $F = \dfrac{s^2_1}{s^2_2}$.

1. The authors thank Donald Y. Young, Department of Ecology, Ethology, and Evolution, University of Illinois, Urbana, Illinois, for contributing this test.

The degrees of freedom for the numerator is $n_1 - 1$, and for the denominator is $n_2 - 1$.

The final step in the F test is to compare the F you have calculated with the critical value for F at the appropriate degrees of freedom (d.f.) using the F table on page 91. If F $_{calculated}$ is greater than F $_{critical}$, then you may reject the H_o. If F $_{calculated}$ is less than F $_{critical}$, you have failed to reject the H_o.

Now back to the t test.

2. The H_o for the t test is that there is no difference between the means of samples 1 and 2. $H_o : \bar{X}_1 = \bar{X}_2$

3. If the variances are unequal, go to step 4. If the variances are equal, calculate t using the formula below: (with d. f. $= n_1 + n_2 - 2$)

$$t = \frac{\left| \bar{X}_1 - \bar{X}_2 \right|}{\sqrt{\dfrac{(n_1 - 1)s^2_1 + (n_2 - 1)s^2_2}{n_1 + n_2 - 2}} \sqrt{\dfrac{1}{n_1} + \dfrac{1}{n_2}}}$$

4. If the variances are unequal, calculate t as follows:

$$t = \frac{\left| \bar{X}_1 - \bar{X}_2 \right|}{\sqrt{\dfrac{s^2_1}{n_1} + \dfrac{s^2_2}{n_2}}}$$

with d.f. $= \dfrac{\left(\dfrac{s^2_1}{n_1} + \dfrac{s^2_2}{n_2} \right)^2}{\dfrac{\left(\dfrac{s^2_1}{n_1} \right)^2}{n_1 + 1} + \dfrac{\left(\dfrac{s^2_2}{n_2} \right)^2}{n_2 + 2}} - 2$

5. When you have calculated your value t, and your d.f., refer to the t table on page 90. At the .05 alpha

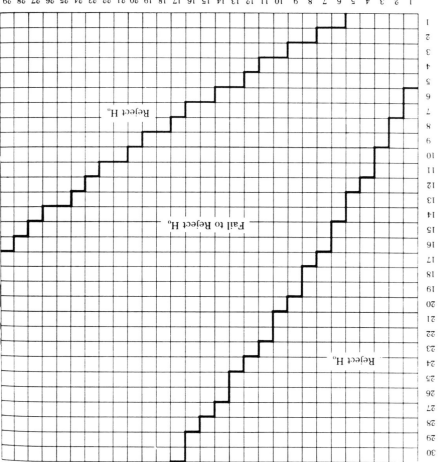

Table for the Sign Test at the alpha level of .05.[1]

1. Constructed from values found in The Chemical Rubber Co., *Handbook of Probability and Statistics*, 2nd ed. (W.H. Beyer, ed.) The Chemical Rubber Company, Cleveland, Ohio. 1968.

level, and at the appropriate degrees of freedom, your critical t value is given in the table. If the t you have calculated in *greater than* the critical value for t in the table, you may reject your null hypothesis. Since you have rejected the null that the means were equal, you may accept your H_1 that they are unequal.

If $t_{calculated}$ is *less than* $t_{critical}$, you have failed to reject your null hypothesis.

Example:

The next table lists compact automobiles, their corresponding country of origin, and their gasoline mileage (compiled a few years ago from *Road and Track* and *Motor Trend* magazines). You would like to know whether the gas mileage of domestic cars differs significantly from that of imported cars.

The parameters calculated from our domestic car data are:

$\bar{X}_d = 19.38 \quad s^2_d = 13.98 \quad n_d = 8 \quad (d = \text{domestic})$

The parameters (miles per gallon) for the imported cars are:

$\bar{X}_i = 26.92 \quad s^2_i = 16.27 \quad n_i = 12 \quad (i = \text{imported})$

In order to select the appropriate t test formula, we must first determine whether or not our calculated variances are significantly different using the F test.

$$H_o : s^2_d = s^2_i \quad F_{calc} = \frac{s^2_i}{s^2_d} = \frac{16.27}{13.98} = 1.16$$

At the .05 alpha level, and d. f. of 11 and 7, we can estimate from our F table (p. 91) that F_{crit} is approximately 4. (If the F table does not accommodate your particular combination of degrees of freedom, you can interpolate.) Since F_{calc} is less than F_{crit}, we fail to reject the H_o and conclude that our variances are not different.

Now that we have established that the variances between the two samples are not different, we can use the t formula given in step three to test our H_o that $\bar{X}_d = \bar{X}_i$

Table of data for the t test example.

Make and Model	Nationality	Miles per gallon
AMC Gremlin	USA	21
AMC Hornet	USA	16
Audi Fox	West Germany	27
Austin Martin	Great Britian	24
Buick Apollo	USA	15
Capri 2600	West Germany	24
Chevrolet Vega	USA	26
Chevrolet Nova	USA	18
Datsun B210	Japan	28
Dodge Colt	Japan	28
Fiat 128	Italy	35
Ford Pinto	USA	23
Ford Maverick	USA	19
Ford Mustang II	USA	17
Honda Civic	Japan	30
Mazda RX-3	Japan	18
Opel Rallye 1900	West Germany	26
Toyota Corolla 1600	Japan	27
Subaru GL	Japan	27
Volkswagen Superbug	West Germany	29

(our H_1 is $\bar{X}_d = \bar{X}_i$), and the formula $n_1 + n_d - 2$ to calculate our degrees of freedom. The degrees of freedom are $12 + 8 - 2$, or 18. When we plug in our parameter values at the appropriate places in the formula for calculating t, it looks like this:

$$t = \frac{26.93 - 19.38}{\sqrt{\frac{(12 - 1)(16.27) + (8 - 1)(13.98)}{12 + 8 - 2}} \sqrt{\frac{1}{12} + \frac{1}{8}}}$$

$t = 4.21$

From the t table (page 90), we find that our t_{crit} at the .05 alpha level with 18 d. f. is 2.101. Since t_{calc} (4.21) is greater than t_{crit}, we may reject our H_0 in favor of our H_1 and conclude that the imported cars have a significantly different gas mileage from that of the domestic cars.

Try this example:

In studying the effects of fatigue on perception, a researcher decides to examine the amount of time that elapses between the first viewing and the visual "reversal" of a Necker cube (an optical illusion). Five students who had been awake for 20 hours were tested and the time to reversal was recorded in seconds. Eight other students who had only been awake for three hours were given the same test and their times recorded.

The data (taken from H. Terrace and S. Parker; see bibliography) looked like this:

Group A awake 20 hrs.	Group B awake 3 hrs.
7.2 sec.	4.7 sec.
5.7	6.3
6.4	4.1
5.3	3.9
6.9	5.4
	5.6
	4.4
	4.8

The null hypothesis here would be that the mean time of reversal in seconds was the same for sample A as for mple B. $H_0 : \bar{X}A = \bar{X}B$

Table of critical values of t for the t Test at the .05 alpha Level.[1]

degrees of freedom	t critical	degrees of freedom	t critical
1	12.706	21	2.080
2	4.303	22	2.074
3	3.182	23	2.069
4	2.776	24	2.064
5	2.571	25	2.060
6	2.447	26	2.056
7	2.365	27	2.052
8	2.306	28	2.048
9	2.262	29	2.045
10	2.228	30	2.042
11	2.201	40	2.021
12	2.179	60	2.000
13	2.160		
14	2.145		
15	2.131		
16	2.120		
17	2.110		
18	2.101		
19	2.093		
20	2.086		

1. Adapted from J.P. Guilford. *Fundamental Statistics in Psychology and Education.* McGraw-Hill Book Company. New York. 1956.

Table of values to be used with the F Test (at the .05 alpha level).[1]

Degrees of freedom in the numerator

	1	2	3	4	5	6	7	8	9	10	12	15	20	24	30	40	60	120	∞
1	647.8	799.5	864.2	899.6	921.8	937.1	948.2	956.7	963.3	968.6	976.7	984.9	993.1	997.2	1001	1006	1010	1014	1018
2	38.51	39.00	39.17	39.25	39.30	39.33	39.36	39.39	39.39	39.40	39.41	39.43	39.45	39.46	39.46	39.47	39.48	39.49	39.50
3	17.44	16.04	15.44	15.10	14.88	14.73	14.62	14.54	14.47	14.42	14.34	14.25	14.17	14.12	14.08	14.04	13.99	13.95	13.90
4	12.22	10.65	9.98	9.60	9.36	9.20	9.07	8.98	8.90	8.84	8.75	8.66	8.56	8.51	8.46	8.41	8.36	8.31	8.26
5	10.01	8.43	7.76	7.39	7.15	6.98	6.85	6.76	6.68	6.62	6.52	6.43	6.33	6.28	6.23	6.18	6.12	6.07	6.02
6	8.81	7.26	6.60	6.23	5.99	5.82	5.70	5.60	5.52	5.46	5.37	5.27	5.17	5.12	5.07	5.01	4.96	4.90	4.85
7	8.07	6.54	5.89	5.52	5.29	5.12	4.99	4.90	4.82	4.76	4.67	4.57	4.47	4.42	4.36	4.31	4.25	4.20	4.14
8	7.57	6.06	5.42	5.05	4.82	4.65	4.53	4.43	4.36	4.30	4.20	4.10	4.00	3.95	3.89	3.84	3.78	3.73	3.67
9	7.21	5.71	5.08	4.72	4.48	4.32	4.20	4.10	4.03	3.96	3.87	3.77	3.67	3.61	3.56	3.51	3.45	3.39	3.33
10	6.94	5.46	4.83	4.47	4.24	4.07	3.95	3.85	3.78	3.72	3.62	3.52	3.42	3.37	3.31	3.26	3.20	3.14	3.08
11	6.72	5.26	4.63	4.28	4.04	3.88	3.76	3.66	3.59	3.53	3.43	3.33	3.23	3.17	3.12	3.06	3.00	2.94	2.88
12	6.55	5.10	4.47	4.12	3.89	3.73	3.61	3.51	3.44	3.37	3.28	3.18	3.07	3.02	2.96	2.91	2.85	2.79	2.72
13	6.41	4.97	4.35	4.00	3.77	3.60	3.48	3.39	3.31	3.25	3.15	3.05	2.95	2.89	2.84	2.78	2.72	2.66	2.60
14	6.30	4.86	4.24	3.89	3.66	3.50	3.38	3.29	3.21	3.15	3.05	2.95	2.84	2.79	2.73	2.67	2.61	2.55	2.49
15	6.20	4.77	4.15	3.80	3.58	3.41	3.29	3.20	3.12	3.06	2.96	2.86	2.76	2.70	2.64	2.59	2.52	2.46	2.40
16	6.12	4.69	4.08	3.73	3.50	3.34	3.22	3.12	3.05	2.99	2.89	2.79	2.68	2.63	2.57	2.51	2.45	2.38	2.32
17	6.04	4.62	4.01	3.66	3.44	3.28	3.16	3.06	2.98	2.92	2.82	2.72	2.62	2.56	2.50	2.44	2.38	2.32	2.25
18	5.98	4.56	3.95	3.61	3.38	3.22	3.10	3.01	2.93	2.87	2.77	2.67	2.56	2.50	2.44	2.38	2.32	2.26	2.19
19	5.92	4.51	3.90	3.56	3.33	3.17	3.05	2.96	2.88	2.82	2.72	2.62	2.51	2.45	2.39	2.33	2.27	2.20	2.13
20	5.87	4.46	3.86	3.51	3.29	3.13	3.01	2.91	2.84	2.77	2.68	2.57	2.46	2.41	2.35	2.29	2.22	2.16	2.09
21	5.83	4.42	3.82	3.48	3.25	3.09	2.97	2.87	2.80	2.73	2.64	2.53	2.42	2.37	2.31	2.25	2.18	2.11	2.04
22	5.79	4.38	3.78	3.44	3.22	3.05	2.93	2.84	2.76	2.70	2.60	2.50	2.39	2.33	2.27	2.21	2.14	2.08	2.00
23	5.75	4.35	3.75	3.41	3.18	3.02	2.90	2.81	2.73	2.67	2.57	2.47	2.36	2.30	2.24	2.18	2.11	2.04	1.97
24	5.72	4.32	3.72	3.38	3.15	2.99	2.87	2.78	2.70	2.64	2.54	2.44	2.33	2.27	2.21	2.15	2.08	2.01	1.94
25	5.69	4.29	3.69	3.35	3.13	2.97	2.85	2.75	2.68	2.61	2.51	2.41	2.30	2.24	2.18	2.12	2.05	1.98	1.91
26	5.66	4.27	3.67	3.33	3.10	2.94	2.82	2.73	2.65	2.59	2.49	2.39	2.28	2.22	2.16	2.09	2.03	1.95	1.88
27	5.63	4.24	3.65	3.31	3.08	2.92	2.80	2.71	2.63	2.57	2.47	2.36	2.25	2.19	2.13	2.07	2.00	1.93	1.85
28	5.61	4.22	3.63	3.29	3.06	2.90	2.78	2.69	2.61	2.55	2.45	2.34	2.23	2.17	2.11	2.05	1.98	1.91	1.83
29	5.59	4.20	3.61	3.27	3.04	2.88	2.76	2.67	2.59	2.53	2.43	2.32	2.21	2.15	2.09	2.03	1.96	1.89	1.81
30	5.57	4.18	3.59	3.25	3.03	2.87	2.75	2.65	2.57	2.51	2.41	2.31	2.20	2.14	2.07	2.01	1.94	1.87	1.79
40	5.42	4.05	3.46	3.13	2.90	2.74	2.62	2.53	2.45	2.39	2.29	2.18	2.07	2.01	1.94	1.88	1.80	1.72	1.64
60	5.29	3.93	3.34	3.01	2.79	2.63	2.51	2.41	2.33	2.27	2.17	2.06	1.94	1.88	1.82	1.74	1.67	1.58	1.48
120	5.15	3.80	3.23	2.89	2.67	2.52	2.39	2.30	2.22	2.16	2.05	1.94	1.82	1.76	1.69	1.61	1.53	1.43	1.31
∞	5.02	3.69	3.12	2.79	2.57	2.41	2.29	2.19	2.11	2.05	1.94	1.83	1.71	1.64	1.57	1.48	1.39	1.27	1.00

Degrees of freedom in the denominator

1. Adapted from R.G.D. Steel and J.H. Torrie. *Principles and Procedures of Statistics.* McGraw-Hill Book Co., New York. 1960.

X² (Chi Square) One-sample Test for Goodness of Fit

Purpose:

This X² test is a test of differences between distributions. It allows one to compare a collection (or frequency distribution) of discrete, nominal data with some theoretical expected distribution to see if they differ significantly. (X is the Greek letter chi, pronounced kī as in "sky.")

Warnings and precautions:

1. The data must be discrete and nominal.
2. When there are only two categories, no expected value may be less than 5. When there are more than two categories, no more than 20% of the expected values may be less than 5, and no expected value may be less than 1.

Null hypothesis:

The observed frequency distribution (or ratio) is equal to the expected frequency distribution (or ratio).

Formula:

$$X^2 = \Sigma \frac{(\text{Obs.} - \text{Exp.})^2}{\text{Exp.}}$$

That is, for the different distributions you are comparing, you square the difference between the observed value and the theoretically expected value, and divide this square by the expected value. The figures calculated for each distribution are then summed for the X².

If the value calculated for X² is *equal to* or *greater than* the critical value given in the table on page 95, for your degrees of freedom, you may reject the null hypothesis. The degrees of freedom (d.f.) for the X² One-Sample Test for Goodness of Fit is equal to the number of categories (or columns — see the example) minus 1.

Example:

You wish to know whether there are more Chevrolets, Fords, or Buicks passing the window of your laboratory during the lunch hour. You count the cars passing from noon to 1:00 pm and find that there are 56 Chevrolets, 71 Fords, and 83 Buicks. For the purposes of your null hypothesis that $56 : 71 : 83 = 1 : 1 : 1$, you need to have a theoretically expected ratio. In this case, the three observed values are summed and divided by three so that you can use your mean as the expected. You would arrange your data in this way:

	Chevrolets	Fords	Buicks	
observed	56	71	83	
expected	70	70	70	
obs. $-$ exp.	-14	1	13	
(obs. $-$ exp.)^2	196	1	169	
$\dfrac{\text{(obs. } - \text{ exp.)}^2}{\text{exp.}}$	2.80	.01	2.41	$X^2 =$ the sum of this line $X^2 = 5.22$

Since there are three categories, your degrees of freedom $= 2$. You check the X^2 table at your appropriate d.f. and find that the critical value is 5.99. Since your calculated X^2 value (5.22) is less than the critical value, you fail to reject the null hypothesis. Therefore you can conclude that there is no statistically significant difference between the number of Chevrolets, Fords, and Buicks passing the window during the hour in question.

Try this example:

Let us say that there are five candidates for political office and a sample of 500 people are interviewed by pollsters to see whom they prefer. In testing the data collected, the null hypothesis would be that there is no difference in preference or that the actual preferences recorded do not differ significantly from 100 people in favor of each candidate. In fact, the results of the interviews look like this:

Candidate	Number of People Preferring the Candidate
JOHN	122
DICK	105
HARRY	86
JIMMY	83
JERRY	104

X^2 Table at the .05 alpha level.[1]

d.f.	critical value
1	3.84
2	5.99
3	7.81
4	9.49
5	11.1
6	12.6
7	14.1
8	15.5
9	16.9
10	18.3
11	19.7
12	21.0
13	22.4
14	23.7
15	25.0
16	26.3
17	27.6
18	28.9
19	30.1
20	31.4

1. Adapted from R.G.D. Steel and J.H. Torrie. *Principles and Procedures of Statistics*. McGraw-Hill Book Co., N.Y. 1960.

X² Test of Independence between Two or More Samples

Purpose:

This X² (Chi square) test, a test of differences between distributions, is used to test for independence between two or more frequency distributions of nominal data. This test is similar in rationale to the X² One-Sample Test for Goodness of Fit. The main differences are that data are put in a contingency table and that the calculations of the expected values are different.

Warnings and precautions:

1. Data must be discrete and nominal.
2. When there are only two categories, no expected value may be less than 5. When there are more than two categories, no more then 20% of the expected values may be less than 5, and no expected value may be less than 1.

The null hypothesis:

The distribution of sample A is equal to the distribution of sample B.

Formula:

$$X^2 = \Sigma \frac{(\text{Obs.} - \text{Exp.})^2}{\text{Exp.}}$$

That is, for the different distributions you are comparing, you square the difference between the observed value and the expected value, and divide this square by the expected value. The figures thus calculated for each distribution are summed to produce the X^2.

If the value calculated for X^2 is *equal to* or *greater than* the critical value given in the table on page 95, for your degrees of freedom, you may reject the null hypothesis. The d.f. for this test = (#columns − 1) (# rows − 1).

Example:

Is the distribution of Fords, Chevys, Buicks, and Opels on a given day the same on Henley Street as on Kingston Pike? In this example, the null hypothesis is:

Henley Street	Kingston Pike
#Fords: #Chevys: #Buicks: #Opels	= #Fords: #Chevys: #Buicks: #Opels

You make your observations and record your data in a contingency table that looks like this:

		Fords	Chevys	Buicks	Opels	row totals	
Henley St.	obs.	45	72	27	90	234	row 1
	exp.	39.8	69.3	65.3	59.6		
Kingston Pk.	obs.	25	50	88	15	178	row 2
	exp.	30.2	52.7	49.7	45.4		
Column totals		70	122	115	105	412 grand total	

Since the theoretical distribution is not readily obvious, we must calculate the expecteds for each cell of the table. You calculate the expected value for each cell by multiplying each cell's row total by its column total and then dividing this product by the grand total. For example, to calculate the upper left expected cell,

$$\frac{234 \times 70}{412} = 39.8$$

The next step of the test is to calculate the

$$\frac{(obs. - exp.)^2}{exp.}$$

for each cell (.68 for the upper left corner). The summation of the above calculations from each cell is the X^2 value. The degrees of freedom for this test = (#columns or categories − 1) times (#rows − 1) or, in this case, 3. $X^2 = 89.66$.

When the X^2 table is consulted (the table is for the .05 alpha level), at the appropriate d.f., we find that the critical value is 7.81. This value is much smaller than our calculated X^2 of 89.66 so we can reject our null hypothesis that the ratios are equal and conclude that the two samples (Henley Street and Kingston Pike) are independent.

Try this example:

One researcher examining the status of mentally deficient adults decided to look for a relationship between being married and level of intelligence. In this case a Chi Square test of independence can be applied to his data to test the null hypothesis that there is no relationship between being married and one's level of intelligence.

The data collected looked like this (data are taken from J.P. Guilford; see bibliography):

Comparison of men of normal IQ with feebleminded men with respect to marital status			
Marital status	Normal	Feebleminded	Both (sample total)
Married	111	84	195
Unmarried	95	122	217
Total	206	206	412

Chapter Eight
Using the Library

by
Diane Schmidt
Assistant Biology Librarian
University of Illinois at Urbana-Champaign

Modern university or college libraries are likely to be much more complicated places than the public or high school libraries that most students are familiar with. The reading rooms or reference areas of most large university libraries are as big as some entire high school libraries, but the same basic organization applies to the largest as well as the smallest libraries. There will be books and journals (though few magazines), and various methods for locating them. Most college and university libraries use the Library of Congress classification system, rather than the Dewey Decimal system used by most school and public libraries, but the underlying principles are the same. Materials on the same subject are grouped together.

In order to find books on a particular subject, you will need to look in a card catalog or in an online catalog, its computer version. (See Fig. 1 for a sample catalog card; most computer versions look similar.) Some libraries have just a single card catalog which you would use to look up books by subject, author's name, or title. Others have separate card catalogs for subject and author or title.

You will be able to find the titles and locations of journals in the card/online catalog, but not individual articles. To find articles, you will need to look up your subject in an index. Indexes are found in either paper form or in electronic form, and your library may have both versions or only one.

The best rule to remember when in the library is that there is nothing to be embarrassed about if you need to ask for help. There is no reason for you to waste your time in fumbling about, and you won't be the only person to ever need help. (On the other hand, don't expect the librarian to do all the work for you!)

Some Definitions

Abstract: a summary of an article, book, etc.

Bibliographic information: information needed to locate a book or article, including author, article title, journal or book title, date, volume, pages, etc. Also called **citation** or **reference**.

CD-ROM: **C**ompact **D**isk-**R**ead **O**nly **M**emory, a data storage format for databases (they look just like your music CDs), usually searched for free.

Database: a collection of information, in the library world usually representing a computerized index.

Index: a detailed list of articles, book chapters, etc., which tells where they are published.

Journal: research or scholarly periodical (more technical than a magazine).

Online catalog: the machine-readable form of the card catalog, used to find books and the location of journals (also known as **OPAC** (**O**nline **P**ublic **A**ccess **C**atalog), **online terminal**, or **public terminal**).

Online database: a database which is available for searching via connection to a remote computer, usually for a fee.

Computers versus paper

More and more finding tools such as indexes and card catalogs are now being put on computers, though books and journals continue to be published mainly in paper form. In many ways, computer-readable indexes and catalogs are better than their paper versions, but there are some drawbacks. It is often much easier to locate material in computer-readable tools than in paper, because there are many more ways to search for items. Each book in the card catalog, for instance, is given only a few subject headings, and you will not be able to locate books if you use the wrong heading. It is not always easy to decide which heading to use (see below for more information). Online catalogs may allow you to do keyword searches, which will locate a word anywhere in the record, such as the book or series title. Another advantage is that printing your results, rather than copying down citations and call numbers, eliminates one major source of error.

One of the major disadvantages of computerized tools is their relative newness. Indexes for articles only cover the time from about 1970 to the present, and many libraries do not offer this entire range. It may only be possible to search for the most recent 2-3 years on a computer, using the paper indexes to look for older material. Some online catalogs only provide access to recent books (again, frequently from about 1970 to the present), and the old card catalog must be consulted for older books. This means that either you must do your search twice, once on computer and once in paper, or only search one source and ignore the older (or newer) material. Once people learn how to use computerized sources, they frequently resist using the old-fashioned indexes, but even in the sciences, old does not mean useless.

Another problem with computerized sources is that there are many different programs used to search them. A card catalog is essentially the same everywhere, but searching an online catalog may vary greatly from library to library. The same is true for indexes: the paper *Biological Abstracts* looks the same in every library, but the computer version may differ. It may be necessary to learn several programs in order to use the computerized tools your library offers, and these programs are subject to change.

Selecting a research topic

It is very important to start with a broad area when selecting a term paper topic, then narrow it down to something that you can write a paper about. Molecular biology, for instance, is much too broad for a five page term paper; it would be far better to focus on a specific area, such as recent advances in the study of ancient DNA. Choosing a narrow topic without knowing something about the broader context can lead to problems when writing your paper.

One good technique for selecting a research topic is to look in a general source such as a textbook or encyclopedia for interesting or significant topics. Another method is to look at current issues of journals, which has the advantage of helping you find topics which are currently "hot." Most libraries keep the current issues of their journals in a separate area, so it is a fairly simple matter to browse a large number of issues.

Another useful, and relatively quick, method which can be used in libraries which have computerized indexes is to use one of them to browse through article citations which were pulled up using a fairly broad subject, such as ecology. (The same can be done in paper indexes, but it is usually faster and easier in computer databases.) Once you find articles which look interesting, you can narrow your topic to a specific area. Browsing through a computer index also allows you to locate articles to use in your paper at the same time as you select your topic.

Locating books and journals

Once you have selected your topic, you need to find books and articles on your subject. If you have started by browsing recent issues or by looking in a computer index, you should already have an article or two in hand. If you look in the references section of the article, you should be able to find some articles on the same subject which the author cited. If you started with a textbook or encyclopedia article, the material cited may be older, but still useful.

In order to find material which is more recent, you will need to use abstracting and indexing periodicals, such as those mentioned in Chapter Nine. Abstracts such as *Biological Abstracts* provide brief summaries of articles, while indexes such as *Biological and Agricultural Index* just provide the basic citation information. Indexes usually appear with less of a time lag than abstracts.

Abstracts or indexes may appear in several formats at your library. In addition to the usual bound volumes of indexes, there may be one or more computerized versions. Your library may have CD-ROMs as well as (or instead of) paper indexes, and some libraries have some indexes mounted on local mainframe computers. These CD-ROMs or local databases can usually be searched for free, but libraries usually have to charge for searching online databases through companies such as DIALOG or BRS. There are many online databases available, and this may be the only way to get access to a particular index. You may not need to use online databases very often, if at all.

To locate books on your subject, you will need to use the card or online catalog. It may be difficult to locate books on your subject, since the headings used are sometimes not what most people expect (movies are listed as moving-pictures, for instance). There may be large red Library of Congress Subject Heading books near the catalog for you to look at. If not, ask a librarian for help. Once you have found one good book, you can check the subject headings for that book to find other headings. If you have an online catalog, use the same strategy or try doing a keyword search if your system allows it.

Another problem with finding books may appear if you try to look for headings that are too specific, or too broad. Books are listed under the most specific heading which can be used for them, so a book solely about spider monkeys would be listed under "spider monkeys," not "primates." On the other hand, your library may not have any books specifically on spider monkeys, so it might be necessary to search broader terms such as "primates" for books which might have a chapter on spider monkeys.

Recording your findings

As you work, be sure to record all of the books and articles consulted. A time-honored and reliable system is to have one 3x5 card for each reference. In addition to the factual information you may need for your paper, record all the bibliographic information you will need for the "literature cited" section. If you head the card with the last name of the first author, the cards will be easy to alphabetize when you are ready to type. Be sure that the information you collect includes the correct name of the author(s), the year of publication, and the title of the article and journal or book title, the volume and issue numbers of the journal, the page numbers of the article or number of pages in the book, and the name and location of the publisher of the book. (See Fig. 2 for examples.) If you are going to quote directly from a book or article, be sure you include the page on which your quotation can be found.

Now that copy machines are so common, many students simply copy the entire article or section of a book which they intend to use; this works well, but make sure that you get the citation information as well. Many journals provide basic citation information on the title page of their articles, but they may not include all the information you need. Copying the title page of a book you use will also be helpful, but again, may not provide all the information you need.

For the technologically advanced, there are many computer databases which can be used to create a bibliography. These are usually called something like personal bibliographic systems and include programs such as ProCite, EndNote and Papyrus. They are useful if you will be doing a lot of research and collecting a lot of articles, since you can enter the information just once and use it to create references for many papers, and the programs will automatically format the references for many different styles. The amount of work needed to create one of these databases may be more than they are worth for an undergraduate, but they are invaluable for anyone doing extensive research.

No matter which note-taking format you use, be sure that your citations are correct and complete. Citations from abstracting and indexing services are usually accurate, but even these services occasionally make mistakes. Citations taken from other papers are often incorrect, and you do not want to perpetuate mistakes which reflect poorly on your research. Likewise, you should always examine the papers you cite, even if you only take information from the abstract or title. Articles which *look* relevant may not be, and this kind of shoddy research is even worse than bibliographic inaccuracies.

Fig. 1: Sample catalog card

The call number identifies the subject of the book and enables you to locate it in the library.

Author

Title

This note gives the number of pages; illustrations, if any; and shows the series title, if any.

Place and date of publication (an indication of how current the information is).

QH	Wilson, Edward Osborne, 1929-
313	The diversity of life / Edward O. Wilson. Cambridge,
W55	Mass.: Belknap Press of Harvard University Press, 1992
1992	424 p. illus. 24 cm. (Questions of science) Includes bibliographic references and index. ISBN 0-674-21298-3

1. Biological diversity. 2. Biological diversity conservation. I. Title. II. Series.

The arabic numerals (tracings) tell which subject headings were used to describe this book. If you look in the catalog under these headings, you can find more books on the same topic.

Fig. 2: Sample citations

Journal article citation

Albert, V. A.; Williams, S. E.; Chase, M. W. Carnivorous plants: phylogeny and structural evolution. Science. 257:1491-1495; 1992.

Book citation

Wilson, E. O. The diversity of life. Cambridge, Mass.: Belknap Press of Harvard University Press; 1992.

Book chapter citation

Marti, C. D.; Korpimaki, E.; Jaksic, F. M. Trophic structure of raptor communities: a three-continent comparison and synthesis. In: Power, D. M., ed. Current Ornithology. New York: Plenum Press; 1993. 10:47-137.

These citations use the format found in the *CBE Style Manual: A Guide for Authors, Editors and Publishers in the Biological Sciences*, 5th ed. Bethesda, MD: Council of Biology Editors, Inc., 1983. Your teacher may want you to use a different format, but you will need the same information as above.

Chapter Nine
An Introduction
To Biological Literature

by
Elisabeth B. Davis
Biology Librarian, University of Illinois
at Urbana-Champaign

Any beginning biology course deserves an introduction to the literature of the field as well as to the subject matter. Unlike the humanities which depend heavily on books to deliver information and ideas, the focus for original research in the sciences is the report issued as a journal article, a conference paper, a technical report, or a patent. The flood of papers, articles, and reports announcing original research and scholarly work brings a difficult problem to the student, since you must find a way to become familiar with the literature, to learn what has happened in the past, and to keep informed about current research.

Over the past twenty years there has been a revolution in the way that information and the flow of scientific literature are managed. Print is giving way to, and in some cases being entirely replaced by, electronic transmission. In the sciences, and especially in biology, the number of journals is mushrooming at an incredible rate requiring faster and more efficient access to the literature of a rapidly expanding body of knowledge.

Since the publication of the first scientific journal in London in 1665, an elaborate system has been developed to assist people in scanning the huge, and ever increasing body of primary research. For ease of discussion and description the scientific literature has been divided into categories describing its function: 1) primary literature reports original work in periodicals, research monographs or reports, patents, dissertations, etc.; 2) secondary literature provides access to primary work using reference materials such as encyclopedias, handbooks, abstracting and indexing serials, bibliographic data-bases, etc.; 3) tertiary literature is *about* science rather than *of* science; good examples are guides to the literature, directo-ries, textbooks, biographies, and the like.

This introduction to the literature of biology is offered in general terms and will concentrate on reference materials that provide entry to the biological literature, in other words, sec-ondary and tertiary sources. Materials are selected to be use-ful at the undergraduate level with the expectation that they will provide access to more technical or comprehensive work, if that is required. Sources are arranged beginning with the more general sources identifying older works, such as ency-clopedias, and progressing to the more specific current aware-ness abstracting and indexing serials. These examples are meant to serve as a guide, as a beginning, not an end to the process of finding information, whether it be in the biological sciences or in some other sphere of knowledge.

Dictionaries and Encyclopedias

Dictionaries and encyclopedias define, discuss, and sum-marize terms and concepts, providing an excellent source for background information on a given topic. Some examples are:

Blinderman, Charles. *Biolexicon: A Guide to the Language of Biology.* Springfield, IL: Thomas, 1990. 363 p. ISBN 0398056714. Bibliography. Index.
Guide to deciphering the vocabulary of biology and medicine.

Cambridge Dictionary of Biology. New York: Cambridge University Press, 1990. 324 p. ISBN 0521397642.
Defines 10,000 terms in zoology, botany, biochemistry, molecular biology, and genetics.

Dictionary of Gardening. New York: Stockton Press, 1992. 4 v. ISBN 1561590010. Also titled *The Royal Horticultural Society Dictionary of Gardening.*
This set includes much more than just gardening: there are biogra-phies of famous botanists, over 180 articles on aspects of plant

biology, glossaries for botany and plant taxonomy, and descriptions for over 50,000 plant genera.

Dictionary of Theoretical Concepts in Biology, compiled by
   K. E. Rowe and R. G. Frederick. Metuchen, NJ:
   Scarecrow Press, 1981. 267 p. ISBN 081081353X.
Provides access to the botanical and zoological literature through 1979 on 1,166 named theoretical concepts. Cites original sources and reviews.

Encyclopedia of Human Biology, Editor-in-Chief R.
   Dulbecco. San Diego, CA: Academic Press, 1991. 8 v.
   ISBN 0122267516 (v.1)
Covers anthropology, behavior, biochemistry, biophysics, cytology, ecology, evolution, genetics, immunology, neurosciences, pharmacology, physiology, toxicology, etc. There is a 40,000 term subject index.

Encyclopedia of Microbiology. J. Lederberg, Editor-in-
   Chief. 4 v. New York: Academic Press, 1992. ISBN
   012226891 (v.1).
Surveys the entire field of microorganisms.

Grzimek's Animal Life Encyclopedia. New York: Van
   Nostrand Reinhold, 1972-1975. 13 v.
Comprehensive encyclopedia to animals, arranged taxonomically. Numerous illustrations and pictures. The section on mammals has been completely revised and is available from McGraw-Hill in five volumes, or on CD-ROM, as *Grzimek's Encyclopedia of Mammals*, 1990. Other companion volumes are *Grzimek's Encyclopedia of Ecology* (Van Nostrand Reinhold, 1976) and *Grzimek's Encyclopedia of Evolution* (Van Nostrand Reinhold, 1976).

Henderson's Dictionary of Biological Terms, edited by E.
   Lawrence. 10th ed. New York: Wiley, 1989. 637 p.
   ISBN 0470214465.
Provides pronunciation, derivation, and definition of terms in biology, from anatomy to zoology.

International Dictionary of Medicine and Biology. New
York: Wiley, 1986. 3 v. ISBN 047101849X.
159,00 definitions cover basic and clinical medical sciences.

McGraw-Hill Encyclopedia of Science and Technology, 7th
ed. New York: McGraw-Hill, 1992. 20 v. ISBN
0079092063 (set).
Comprehensive science and technology encyclopedia.

*Merck Index: An Encyclopedia of Chemicals, Drugs, and
Biologicals,* 11th ed. Rahway, NJ: Merck, 1989. 1 v.
ISBN 091191028X. Available online from BRS and
DIALOG.
Defines and describes more than 10,000 significant drugs, chemi-
cals, and biologicals.

*Official World Wildlife Fund Guide to Endangered Species
of North America.* Washington, DC: Beecham
Publishing, 1990-1992. 3 v.
Comprehensive descriptions, including photographs, of plants and
animals that are listed as either endangered or threatened.

Handbooks

Handbooks, usually consisting of summary information in
the form of charts, graphs, tables, and the like, are designed
to be one volume reference books. Field guides and classification schemes are good examples of the sort of handbook
that you may find useful. There are also some general biological handbooks listed as examples here, but if you need
more specialized information, consult a detailed guide to the
literature of the topic. As an illustration, if you need more
information about sources for entomology, see Gilbert and
Hamilton's *Entomology* listed in this section. Other examples
of handbooks include:

Bergey's Manual of Systematic Bacteriology. 4 v. Balti-
more: Williams & Wilkins, 1984. ISBN
0683041088 (v. 1).
Diagnostic keys and tables for identification; detailed descriptive
information and taxonomic comments.

Beynon, R. J. *Postgraduate Study in the Biological Sciences: A Researcher's Companion.* Brookfield, VT: Portland Press, 1993. 150 p. ISBN 1855780097.
Basic advice and information for postgraduate study, experimental skills, research projects, scientific writing, public speaking, computers, teaching, beginning career moves.

Biosciences: Information Sources and Services, edited by Y. Alston and J. Coombs. New York: Stockton Press, 1992. 407 p. ISBN 1561590479.
This began as a directory for biotechnology but has expanded to include activities, products, and developments in fundamental and applied biosciences.

CBE Style Manual: A Guide for Authors, Editors, and Publishers in the Biological Sciences, 5th ed, rev. and expanded. Bethesda, MD: Council of Biology Editors, 1983. 324 p. ISBN 0914340042.
The standard for style in the biological sciences.

CRC Handbook of Chemistry and Physics, latest ed. Boca Raton, FL: CRC/Lewis, 1994.
Annual publication that is essential for information on chemical, physical, and engineering data. Includes mathematical tables and other information necessary for biologists.

Davis, E. B. and D. Schmidt. *Using the Biological Literature; A Practical Guide,* 2nd ed. New York: Marcel Dekker, 1995.
A comprehensive guide to the biological literature.

Field Guides. These are usually regional and/or plant and animal specific. There are several excellent field guide series that are particularly useful. For example, look for the *Audubon, Golden,* or *Peterson Field Guides* in your library or bookstore.

Five Kingdoms: An Illustrated Guide to the Phyla of Life on Earth, 2nd ed. New York: Freeman, 1987. 376 p. ISBN 0716719126.
Discusses classification schemes and general features of each kingdom, including illustrations and a bibliography for additional information.

Gilbert, P. and C. J. Hamilton. *Entomology; A Guide to Information Sources*, 2nd ed. New York: Mansell, 1990. ISBN 0720120527.
Introduction and source book for entomology.

Illustrating Science. Standards for Publication. Scientific Illustration Committee, Council of Biology Editors. Bethesda, MD: CBE, 1988. 296 p. ISBN 0685218937.
Standards and guidelines for publication.

Synopsis and Classification of Living Organisms. S. P. Parker, Editor-in-Chief. 2 v. New York: McGraw-Hill, 1982. ISBN 0070790310 (set)
Classification of living organisms with descriptions and 8,200 summary articles for all taxa down to the family level.

Information Sources in the Life Sciences, 3rd ed. Edited by H. V. Wyatt. London: Butterworths, 1987. 191 p. ISBN 040811472X
Guide to the literature.

Reviews of the Literature

Reviews of the literature are exactly that: State-of-the-art reports of research, techniques, and methods covering a specific period of time, for a particular subject. Review series are easy to spot because their titles often begin with *Advances in...*, *Annual Review of...*, *Current Topics in...*, *Progress in...*, *Trends in ...*, *Yearbook of....* Reviews may provide background information, may indicate classic papers, and may point out directions for future research. Whatever their aim, there is always a lengthy bibliography of citations to the literature included in the review.

Reviews to the literature are indexed in the serial publication *Index to Scientific Reviews*, and they are indexed in the general biological abstracting and indexing serials that are annotated in the **Current Awareness** section of this chapter. General biological review articles may also be found in the magazine *Scientific American* and the *Quarterly Review of Biology*.

Treatises

Treatises, usually requiring several volumes, may present original research, a compilation of information, a thorough review of a subject, or a combination of all three. Early work in the field and classic experiments are often discussed with an attempt to examine the subject in a comprehensive fashion. You can locate specific titles for books of this sort by using guides to the literature mentioned earlier, by consulting retrospective or current bibliographies, or by using *Science Citation Index*. Some examples of well-known treatises are:

The Bacteria; A Treatise on Structure and Function, edited by I.C. Gunsalus et al. Multivolume. New York: Academic Press, 1960- .

Handbook of Physiology; A Critical, Comprehensive Presentation of Physiological Knowledge and Concepts. Multivolume. Bethesda, MD: American Physiological Society, 1959- .

Hutchinson, G. E. *A Treatise on Limnology*. 4 v. New York: Wiley, 1957-1993.

Wright, Sewall. *Evolution and the Genetics of Populations; A Treatise*. 4 v. Chicago: University of Chicago Press, 1968-1978.

Basic Current Awareness Resources

When your review search of the literature has been finished, the next step is to focus on current research. Original work appears in primary sources, as articles in journals. To get an idea of the scope of the biological literature, take a look at the *Serial Sources for the BIOSIS Previews Database*. This serial list is published annually by the BIOSIS people in Philadelphia and includes complete name, abbreviation, and publishing information for over 9,000 primary biological journals. Obviously, scanning the journal literature looking for relevant articles could be a daunting task. However, it can be managed most efficiently and effectively by using established abstracting and indexing serials to find information. All of the general biological abstracting and indexing sources are currently available in several versions: print, computerized database, CD-ROM, or magnetic tape. In the past, abstracting indexes had the longest lagtime between appearance in a journal and being indexed; it took the most current index six to eight weeks to report the appearance of an article. In the current electronic environment indexes can be composed much more rapidly with almost instantaneous dissemination. Following are short descriptions for some of the most important abstracting and indexing serials for biology:

Bibliography of Agriculture. v.1- , 1942- . Phoenix:
Oryx. ISSN 0006-1530.
Scans international agricultural literature in journals, government documents, reports, and proceedings compiled by the National Agricultural Library. Available in print, online, CD-ROM and magnetic tapes.

Biological Abstracts. v. 1- , 1926- . Philadelphia: BIOSIS.
ISSN 0006-3169.
The most comprehensive biological abstracting service in the English language in the world. *BA* scans over 9,000 primary journals and covers all of biology including biomedicine. Available in print, as a computerized online database, CD-ROM, and magnetic tapes.

Biological Abstracts RRM (Reports, Reviews, Meetings). v.
    18- , 1980- . Philadelphia: BIOSIS. ISSN 0192-
    6985.
This companion to *BA* covers original work not reported in *BA*,
specifically, editorials, bibliographies, proceedings, symposia,
books, nomenclature rules, etc. Availability is similar to *BA*.

Biological and Agricultural Index. v.1- , 1916/18- . New
    York: Wilson. ISSN 0006-3177.
Useful for beginning students and the public. Covers 225 journals.
Available in print and magnetic tapes, online from BRS, and on
CD-ROM from Wilson.

Biology Digest. v.1- , 1974- . Medford, NJ: Plexus. ISSN
    0095-2958.
Covers 200 biology journals and is appropriate at the undergraduate level and for the general public. Available in print and online on
the Life Sciences Network.

Chemical Abstracts. v.1- , 1907- . Columbus, OH: Chemi-
    cal Abstracts Service. ISSN 0009-2258.
The most important English language index for chemistry. Covers
18,000 journals. Available in print, online, and on CD-ROM.

Current Awareness in Biological Sciences. v.1 -, 1954- .
    Tarrytown, NY: Pergamon. ISSN 0733-4443.
Covers 2,700 journals. Available in print, and online.

Current Contents/Life Sciences. v.1- , 1958- . Philadelphia:
    Institute for Scientific Information. ISSN 0011-
    3409.
Compilation of tables of contents for 1200 life sciences journals;
also available in a 600 title version. Includes title word subject and
author indexes as well publisher's addresses. The companion
publication *Current Contents/Agriculture, Biology, and Environmental Sciences* is another section of the Institute for Scientific
Information database that is pertinent to biology, covers different
journals, and is available, like *CC/LS*, in print, online, CD-ROM,
and magnetic tapes.

Dissertation Abstracts International. v.1- , 1938- . Ann
Arbor, MI: University Microfilms International,
Dissertation Publishing. ISSN 0319-4217.
Covers U. S. and Canadian academic institutions; British and
European dissertations are included since 1988. Available in print,
online, and CD-ROM.

General Science Index. v.1- ,1978- . New York: Wilson.
ISSN 0162-1963.
Scans 139 magazines and journals in all sciences. Appropriate for
non-specialists. Availability is the same as *Biological and Agricultural Index.*

Government Reports Announcements and Index. v. 75- ,
1975- Springfield, VA: U.S. Department of Commerce, National Technical Information Services.
ISSN 0097-9007.
Lists technical reports and research supported by federal grants and
some state and local governments. Available in print, online, and
on CD-ROM.

Index Medicus. New series, v.1- , 1960- . Washington, DC:
National Library of Medicine. ISSN 0019-3879.
The most comprehensive international medical database. Available
in print, magnetic tapes, online as *MEDLINE*, and on CD-ROM.

Life Sciences Collection. Bethesda, MD: Cambridge Scientific Abstracts.
Computerized index to over 5,000 journals. Available in print in
several abstracting subject sections, online, and on CD-ROM. A
chief competitor to *Biological Abstracts* although the coverage for
botany is not as strong as *BA*.

Monthly Catalog of United States Government Publications. v.1- , 1985- . Washington, DC: Government
Printing Office. ISSN 0362-6830.
Essential for locating U. S. government documents. In print,
online, and on CD-ROM.

Official Gazette of the United States Patent and Trademark Office. Patents. v. 1- , 1872- . ISSN 0098-1133.
Patents with abstracts and sketches are listed weekly. In print, online, and on CD-ROM.

Science Citation Index. v.1- , 1961- . Philadelphia: Institute for Scientific Information. ISSN 0036-827X.
Covers articles in 3,000 international journals for all the sciences. Unique index provides access to classic papers and cited papers. Available in print, online, CD-ROM, and magnetic tapes.

Zoological Record. v.1- , 1864- . Philadelphia: BIOSIS. ISSN 0144-3607.
Comprehensive international zoological index with exhaustive coverage for the systematic literature. Covers 6,500 journals plus books and proceedings. Available in print, online, and on CD-ROM.

Chapter Ten
How to Write a Scientific Paper

The organization of a research paper reflects the basic pattern of research design. Scientific papers follow a rigid format which is extremely helpful to both writer and reader. Although the highly structured format will be new to many of you, you may find that writing a scientific paper is easier than writing a paper for a humanities course. Cleverness, beauty, originality, and style (although extremely important in the design of experiments) are not required in scientific writing. What is required is a clear, logical, orderly presentation of what your question was, how you planned to answer it, what your results were, and what you conclude. Needless to say, good grammar and precise wording are crucial to effective communication.

The scientific paper has the following elements: Title, Abstract (or sometimes a final summary instead), Introduction, Methods and Materials, Results, Discussion, and Literature Cited. The actual words *Introduction, Methods and Materials,* etc. are used to head the sections of your paper. These words are centered on their line and followed by the text for that section. Normally you do not begin a new page for a new section unless the preceding section completely filled the page. Illustrative tables and figures may be inserted where appropriate or placed at the end of the paper. These items must be fully labelled (not just "Figure 1" but an explanation of what is being shown — if it is a graph the axes must be named and the units of measurement given, for example) so that they would be understood by someone who had not yet read the paper. Data or experimental results presented in a graph or table must also be summarized verbally in the text so that the text could be understood by someone who had not seen the table.

In this chapter we will illustrate the sections of the scientific paper using examples from actual published material and also from unedited papers written by students taking an introductory biology course taught in the investigatory laboratory mode.

I. The Title.

The title of a scientific paper should tell the reader what kind of work is being reported. If possible, it should reveal the organism studied, the particular aspect or system examined, and the variable(s) manipulated. A common student failing is to give a paper an uninformative title such as "Ecology Experiment." Although titles should be simple, direct, and informative, there is room for some grammatical variation. Here are some sample titles which are appropriate for their purposes.

THE EFFECT OF TEMPERATURE ON THE RATE OF HEARTBEAT IN *DAPHNIA*

DOES TEMPERATURE AFFECT THE GERMINATION RATE OF CORN?

FRECKLES AND HAIR COLOR: A SEARCH FOR SEX-LINKAGE

WEB-SITE SELECTION IN THE DESERT SPIDER *AGELENOPSIS APERTA*

A NEW SPECIES OF LIZARD OF THE GENUS *AMEIVA* (TEIIDAE) FROM THE PACIFIC LOWLANDS OF COLOMBIA

COMPARATIVE ASPECTS OF INVERTEBRATE EPITHELIAL TRANSPORT

SEASONAL EFFECTS OF DEHYDRATION IN AIR ON UREA PRODUCTION IN THE FROG *RANA PIPIENS*

CONTRACTION OF SINGLE SMOOTH MUSCLE CELLS FROM *BUFO MARINUS* STOMACH

POLARIZING ACTIVITY IN THE DEVELOPING LIMB OF THE SYRIAN HAMSTER

II. The Abstract

The Abstract is a one or two paragraph condensation of the entire article giving the main features and results of the work described more completely in the article. It helps the reader decide whether the material in the paper will be worth reading. Abstracts are often published separately from the paper in literature searching services such as *Biological Abstracts* (see Chapter Nine). Reducing a long paper to a few paragraphs is an art and naturally requires practice. Here are some sample abstracts — three from published reports and two from student papers.

THE COLOR PATTERN OF *SONORA MICHOACANENSIS*
(DUGES) (SERPENTES, COLUBRIDAE) AND ITS
BEARING ON THE ORIGIN OF THE SPECIES
by Arthur C. Echternacht
Breviora, 1973. 410:1-18

Abstract

The extensive variation in color pattern of the 31 known specimens of *Sonora michoacanensis* is described and a model illustrating the relationships of the major components presented. *Sonora aequalis* Smith and Taylor is placed in the synonymy of *Sonora michoacanensis mutabilis* Stickel from which it differs only slightly in color pattern. It is suggested that *S. michoacanensis* evolved from a bicolor, banded ancestor within the *S. semiannulata* group or from a common ancestor at the southern edge of the Mexican Plateau following habitat shifts associated with climatic changes during the Pleistocene. *Sonora michoacanensis* is interpreted as an imperfect Batesian mimic of elapid coral snakes (*Micrurus* sp.), intermediate in an evolutionary sequence beginning with the bicolor, banded ancestor and leading toward a more perfect, tricolor mimic. Known locality records of *S. michoacanensis* are mapped and selected merisitc data presented in tabular form.

EVIDENCE FOR A GRADIENT OF A MORPHOGENETIC SUBSTANCE IN THE DEVELOPING LIMB

by Jeffrey A. MacCabe and Brenda W. Parker

Developmental Biology, 1976. 54:297-303

Abstract

A polarizing activity in the developing vertebrate limb bud has been implicated in the control of the morphogenesis of its anteroposterior axis. The results of an earlier experiment suggest that there is a gradient of a similar morphogenetic activity in the 4-day chick wing, with a high level of activity along the posterior border, no detectable activity at the anterior border, and an intermediate level of activity in the center, i.e., about halfway between the anterior and posterior borders. The experiments reported here show that this activity disappears from the center of the limb after placing an impermeable barrier posterior to the center, but not after placing the barrier anterior to the center. A porous barrier placed posterior to the center does not alter the activity in the center of the limb. These results lead us to suggest that the gradient of morphogenetic activity in the chick wing is the result of the movement (possibly by diffusion) of a factor(s) from a source in the posterior region of the wing.

PREY PREFERENCE AND HUNTING HABITAT SELECTION IN THE BARN OWL

Stephen J. Fast and Harrison W. Ambrose III

American Midland Naturalist. 1976. 96 (2):503-507

Abstract

A series of experimental situations permitting a barn owl (*Tyto alba*) various combinations of choices between woods-like and field-like habitats, and between *Peromyscus leucopus* and *Microtus pennsylvanicus* as prey items, demonstrated that the owl has statistically significant preferences for *Microtus* and the field hunting habitat.

COMPARATIVE CHEMICAL ANALYSIS OF WATER FROM TWO PONDS OF DIFFERENT AGES

by Michael Brown and Judy Conley

(students — University of Illinois)

Abstract

Two man-made lakes differing in age by about five years (the older one located north of Champaign, Illinois at the junction of Interstate 74 and Interstate 57, and the younger one located just east of LeRoy, Illinois on Interstate 74) were compared on a chemical basis. Spectrophotometric assays were performed for nitrate, sulfate, phosphorus, lead, and anionic surfactants. A significant difference was found in the mean sulfate and phosphorus concentrations. Environmental implications of this finding and predictions about the aquatic flora are discussed.

THE EFFECT OF VARIOUS LIGHT CONDITIONS
ON MATING OF *DROSOPHILA MELANOGASTER*
by James F. Palma and Carole Cunningham
(students — University of Illinois)

Abstract
The purpose of our experiment was to determine whether light conditions have any effect on the mating of fruit flies. We obtained virgin fruit flies and placed males and females in agar-filled test tubes. We placed the test tubes under bright, normal, and dim light conditions and observed the flies to see if mating took place in any or all of these conditions. We found that flies had no contact in the normal and darkened conditions, while in the bright light, all the flies mated within two hours.

III. The Introduction

The introduction should present the question being asked and place it in the context of what is already known about the topic. Background information which suggests why the topic is of interest and related findings by other scientists are usually mentioned here. (The format for citing other works will be discussed in Chapter Ten.) Some sample introductions from published material and from student papers follow.

THE ROLE OF STORED GLYCOGEN DURING
LONG-TERM TEMPERATURE ACCLIMATION IN THE
FRESHWATER CRAYFISH, *ORCONECTES VIRILIS*
by Arthur M. Jungreis
Comp. Biochem. Physiol. 1968. 24:1-6.

Introduction
In Crustacea, glycogen metabolism has been associated with chitin synthesis (Renaud, 1949; Passano, 1960). However, the role played by glycogen in intermediary metabolism

under normal conditions or during starvation is not clear. According to Renaud, crustacean intermediary metabolism centers around glycogen and fatty acids, while according to Scheer and co-workers (Scheer & Scheer, 1951; Kincaid & Scheer, 1952; Scheer et al., 1952; Neiland & Scheer, 1953) the primary energy source is protein and not carbohydrate and fat. When ^{14}C-labeled glucose was injected into the spiny lobster *Panulirus*, it appeared almost exclusively as labeled glycogen rather than labeled CO_2 (Scheer & Scheer, 1951). Furthermore, during an artificial period of starvation, the total glycogen content in *Panulirus* did not decrease (Scheer & Scheer, 1951).

To determine the role of glycogen as a metabolic source of energy during acclimation in long-term starved crayfish *Orconectes virilis*, glycogen content was analyzed in a variety of tissues after 45 days of acclimation.

CYCLIC CHANGES IN THE ALCOHOL-SOLUBLE CARBOHYDRATES IN SYNCHRONIZED *TETRAHYMENA*
by G.L. Whitson, J.G. Green, A.A. Francis, and D.D. Willis
Journal of Cellular Physiology, 1967. 70:169-177

Introduction

Cyclic changes and the sequential synthesis of macromolecules in cells are of particular interest to cell biologists involved in studies of cell division. In most cases, direct biochemical and biophysical analyses are limited when large quantities of synchronized cells are not available. Small populations synchronized by picking out individual dividing cells have been effectively used in cytological, autoradiographic, and interference-microscope studies concerning the relations of DNA, RNA, protein, and dry-weight changes to cell division in both *Paramecium* (Kimball and Barka, 1959; Kimball et al., 1959; Kimball and Perdue, 1966) and *Tetrahymena* (Prescott, 1960, 1964). Temperature changes used to induce synchronization in *Tetrahymena* (Scherbaum and Zeuthen, 1954; Padilla and Cameron, 1964) have made possible the use of large quantities of cells for direct biochemical and biophysical studies on macromolecular changes in this organism (Zeuthen, 1964; Padilla et al., 1966; Whitson et al., 1966a,b).

The lack of information concerning the role of soluble carbohydrates in cell division in *Tetrahymena* prompted this investigation. Although the role of the carbohydrates in energy metabolism in *Tetrahymena* is well understood (Kidder and

Dewey, 1951), very little information has been reported about either the changes in content or the identification of soluble carbohydrates in this organism (Kamiya, 1959; Chou and Scherbaum, 1965a, b). Kamiya (1959) has reported that the total alcohol-soluble sugar content in *Tetrahymena* decreases during synchronization and fails to rise after cell division. He observed more marked changes in the acid-soluble fractions where hexose and pentose sugars increased during the synchronizing treatment, were highest in concentration before cell division, and then decreased.

In contrast, we have found that 15 alcohol-soluble carbohydrate compounds appear in cell extract of *Tetrahymena* during the predivision period after synchronization. The concentrations of most of these compounds increase to a maximum prior to cell division, begin to decrease during cell division, and are lower after cell division occurs. Attempts to show changes in acid-soluble fractions were not pursued when it was found that very little extractable carbohydrate material was obtained with repeated TCA extractions. Changes in the alcohol-soluble carbohydrates were recorded with a prototype automated carbohydrate analyzer (Green, 1966) which employs the elution of sugars as borate complexes (Khym and Zill, 1952).

THERMAL BALANCE AND PREY AVAILABILITY: BASES FOR A MODEL RELATING WEB-SITE CHARACTERISTICS TO SPIDER REPRODUCTIVE SUCCESS
by Susan E. Riechert and C. Richard Tracy
Ecology, 1975. 56 (2):265-284

Introduction

A definite pattern observed in the local distribution of animals can signify the underlying adaptations of individuals to their physical environment. Within a diverse group, those individuals that find suitable locations not only increase their chances of survival to maturity, but also are likely to contribute the greatest number of offspring to the next generation, i.e., exhibit greater fitness (Williams, 1966; Pianka, 1974). Therefore, the genotypes of individuals demonstrating greater habitat discrimination will predominate. This adaptation by the majority of the individuals to selection of favorable habitat is reflected in the local pattern of the population.

Agelenopsis aperta (Gertsch) is a member of the funnel web-building spider family, Agelenidae (Araneae). The web consists of a flat sheet with an attached funnel extending into some feature of the surrounding habitat. Occasionally a scaffolding is present. The sheet has no adhesive properties and serves merely as an extension of the spider's legs. *Agelenopsis* carries out much of its activity within a sheltered environment, coming out of the funnel only as long as required for securing prey and repairing the web. Study of the local distribution of this spider has demonstrated the presence of patterns related to three functions: social, reproductive, and vectorial (Riechert et al., 1973; Riechert, 1974b). A vectorial pattern is influenced by factors of the external environment (e.g., gradients of temperature and humidity, Hutchinson, 1970) and in this context is observed in the association of spiders with specific habitat features. For the funnel web spider, at least, this association does not result from differential survival of randomly dispersed individuals, but rather reflects the active selection of specific web-site characteristics by the spiders (Riechert, 1973). We postulate that individuals, in selecting certain habitat characteristics, are more fit than those showing less, or erroneous discrimination. In this paper, we assess the relationship between the presence of various web-site characteristics and the reproductive success of individuals at web sites offering these characteristics.

THE RELATIONSHIP OF CAFFEINE TO CHROMOSOMAL BREAKAGE IN ONION ROOTS
by Pam Silvers and Bob Montiz
(students — University of Illinois)

Introduction
Dependency on caffeine has become a typical American syndrome. Caffeine is present in a majority of our popular beverages (coffee, tea, Coca-Cola, etc.) which are consumed in large quantities. Does anyone ever stop to think about possible consequences? There has been some evidence presented that caffeine causes chromosomal breakage and inhibits cytokinesis (Briesele, 1958). This experiment was designed to examine the effects of caffeine, at varied concentrations and time intervals, on the chromosomes of the onion root tip.

A COMPARISON OF SOIL ARTHROPOD SPECIES
DIVERSITY AND PLANT SPECIES DIVERSITY IN
SELECTED SAMPLES
by Evelyn Mueller and Paula Prindle
(students — University of Illinois)

Introduction

The diversity of life on earth has long astounded man. His relationship to other organisms, and the interactions among those organisms, have often been keys to common survival. "Each organism on earth is a member of an ecosystem, a unit that consists of the other organisms that affect it" (Wilson, 1974). One of the most fascinating aspects in ecology, the study of the interactions of organisms with their environment, is the concept of species diversity. Our investigation sought to compare the number of species of plant life on a soil sample to the number of species of soil arthropods extracted from the soil underneath.

IV. Methods and Materials

This section of the paper should describe the materials and procedures used in sufficient detail that others could repeat the research if they wanted to see if they could get similar results. Specific, published techniques can be cited without being described in detail. The method of approach to the problem should be narrated in the past tense, telling what *was done* not telling what the reader *should do*. Diagrams of the experimental apparatus or study areas are often helpful. Because this is often a very long portion of the paper, in some cases excerpts will be given rather than entire sections.

GRADED CONTRACTIONS IN MUSCLE STRIPS AND
SINGLE CELLS FROM *BUFO MARINUS* STOMACH
by Roland M. Bagby and Bruce A. Fisher
American Journal of Physiology, 1973. 225:105-109

Methods and Materials

Adult specimens of *Bufo marinus* (200-300g) were double pithed and the stomachs were removed. Single smooth muscle cells were prepared according to Bagby et al. (1971).

For preparing strips, stomachs were stretched over tapered

glass rods, the serosa was peeled away, and strips of muscle 2 mm wide were separated from the underlying mucosa. Microscope observations of sections cut from strips prepared in this manner showed that the strips contained only the circular smooth muscle layer. The strips were mounted vertically in a 100-ml bath filled with a modified Harris-Ringer solution (Harris) of the following composition, in grams per liter: NaCl, 5.2; KCl, 0.22; NaHCO$_3$, 2.52; Na$_2$HPO$_4$·12H$_2$O, O.83; CaCl$_2$, 0.28; MgSO$_4$, 0.12; and glucose, 0.54. The upper end of a strip was attached to the moving element of a LVDT-type of isotonic transducer (Narco Bio-Systems, Inc., MkII) which had been modified to weigh only 225 mg. The weight of the element and attached muscle could be counter-balanced to produce the desired load on the muscle. In this investigation, a load of 100 mg was used for muscle strips 1.5-2.0 mm in diameter. d-c Pulses from a Grass S44 stimulator and SIU5 stimulus-isolation unit were applied transmurally by platinum wire-loop electrodes surrounding the strips. Responses were recorded on a Physiograph DMP-4 polygraph.

Single smooth muscle cells were pipetted onto slides having two silvered strips 6 mm apart which were connected to a Grass S44 stimulator and SIU5 stimulus-isolation unit, allowing a d-c field stimulus to be passed through the droplet containing the cells. Responses were recorded on 16 mm Tri-X reversal film with a Bolex movie camera and Wild phototube mounted on a Nikon Suke microscope with phase-contrast optics. The time of stimulus was marked when stimulus was applied. A negative for making prints was obtained by making a contact print of the original on Eastman fine-grain release positive film.

WEB-SITE SELECTION IN THE DESERT SPIDER
AGELENOPSIS APERTA
by S.E. Riechert
Oikos, 1976. 27:311-315

Methods and Materials

The study was conducted at the northern edge of the Malpais Lava Beds, Lincoln Co., New Mexico (Robert M. Shafer Ranch, T6S R1OE. 1636m). Descriptions of the vegetation characteristic of the lava and desert grassland habitats and of the life cycle of *A. aperta* in South Central New Mexico are given elsewhere (Riechert et al., 1973; Riechert, 1974).

Individuals within desert grassland and lava bed study areas were marked in 1971 and 1972 in the following manner. Spiders were captured, etherized, sexed, and paint-marked on the dorsum of the abdomen with one or more colors of a non-toxic, fast drying enamel paint. A leg removal technique was used in immature individuals where paint would be lost upon molting; regenerated legs could be readily distinguished. A total of 222 individuals were marked during the course of the study. The exact locations of occupied webs were mapped and environmental characteristics of web-sites were determined from line-intercept sampling techniques. At each site a line intercept 1 m in length was run in a north-south compass direction bisecting the center of the web sheet with the 50 cm mark positioned at sheet center. Presence and height of plant species, depression depth, and presence of various substrate features were recorded at 10 cm intervals. An additional transect was sampled at a regular distance away from each web (beginning 1 m west of the end of the web transect). This second transect was defined as a nonweb transect regardless of its potential for a past or future web.

The presence or absence of spiders was determined by flushing them from their web funnels each morning between 0800 and 1000, MDST, coinciding with maximum activity (Riechert and Tracy, 1975). Individuals were scored as flushed if they appeared at the funnel entrance when their webs were approached. If a spider did not flush, the web was studied for signs of disuse (i.e., a torn or littered web) and the surrounding area was checked for a new web. When located, new webs were mapped and the occupants flushed to determine their marked status.

THE EFFECTS OF ACTINOMYCIN D AND RIBONUCLEASE
ON ORAL REGENERATION IN *STENTOR COERULEUS*
by G.L. Whitson

Journal of Experimental Zoology, 1965. 160 (2):207-214

Methods and Materials

Culturing and cutting techniques. Cultures of *Stentor coeruleus* were obtained from the North Carolina Biological Supply Company. Subcultures were maintained at room temperatures (23-25°C) on baked lettuce infusion pH 6.8-7.2 with *Tetrahymena* sp. and *Aerobacter aerogenes* as food organisms.

Cells used for experimental analysis were grown in deep Petri dishes and small depression slides. Refeeding of the stentors was done by daily addition of *Tetrahymena* taken from cultures grown in lettuce infusion which were innoculated daily with fresh bacteria. Both mass cultures and individual isolation lines were maintained during most of this investigation.

Cells for regeneration experiments were pipetted in a large drop of culture fluid onto a small piece of fine-mesh cotton cloth which was attached to a slide with melted paraffin. They were then cut with flame-sharpened tungsten needles kindly provided by S.K. Brahma. A group of tail pieces was obtained by cutting cells during a 15-30 minute interval and placing them in fresh culture medium in spot-plate depressions. In this way it is possible to observe a group of 50-60 tail pieces that will ordinarily regenerate at approximately the same rate.

The use of inhibitors. Actinomycin D (Merck Sharp and Dohme Research Laboratories, West Point, Pa.) and crystalline ribonuclease (Worthington Biochemical Corporation, Freehold, N.J.) were used as inhibitors of regeneration. Several concentrations of actinomycin were made up in autoclaved lettuce infusion adjusted to pH 7.0 with 1NNaOH. Ribonuclease in different concentrations was also prepared, and used in either autoclaved lettuce infusion or distilled water adjusted to pH 7.0 with phosphate buffer. Tests with both of these inhibitors were performed on whole cells prior to cutting and on isolated tail pieces at different times during the course of regeneration. Both starved and well-fed cells were used. The results were recorded as no inhibition, partial inhibition, or complete inhibition of regeneration.

MOVEMENT RESPONSES OF FEMALE *ACHETA DOMESTICA* TO SELECTED SOUNDS

by Evelyn Mueller
(student — University of Illinois)

Methods and Materials

A cardboard experimental box was constructed of the following approximate dimensions: 91.44 cm long by 30.48 cm wide by 45.72 cm high (see figure 1). Within this box, an inner runway was sectioned off by taping 4 walls of cardboard of the same height as the box (45.72 cm) to the floor of the box at a distance of 7.62 cm in from the outer edges. The tape formed a seam between the bottom of the box and the new inner walls, so that the crickets could not escape from the inner runway. The floor of the inner runway was marked into eight 7.62 cm-wide strips numbered consecutively 1 through 8. Newspaper was stuffed between the outer box wall and the inner compartment wall as soundproofing material. Two spaces of about 7.62 cm in width were left open at the ends (designated A and B in the figure) where the cassette tape recorders were to be inserted. Once the cassette recorders were placed in the designated slots, their speakers touched the end walls of the inner compartment.

. . . Five trials were performed. 20 female *Acheta domesticus* crickets were used in each of the five trials. The first trial was a test for the box bias. Tape recorders were placed in their designated slots, but no sound was emitted from either. The crickets were released in a silent box.

In the second trial, the pure tone recording, "1" was played on the cassette recorder at one end of the box and the cassette recorder at the opposite end was not played.

The third trial employed the recording of the male-female encounter, tape "2." I interpreted this song to be a combination of the calling and courtship songs of *Acheta domesticus* (see figure 3). Tape 2 was played from the recorder at box end "A." The cassette recorder at end "B" was not played . . . . . .

SEX LINKAGE IN *DROSOPHILA MELANOGASTER*
by Cary Miller and Lynn Johnson
(students — University of Illinois)

Methods and Materials

To begin our experiment, we obtained two vials of fruit flies. One contained homozygous wild type (+) male and female fruit flies while the other contained homozygous white apricot-eyed (*wa*) male and female flies. These flies were known to be homozygous because neither vial exhibited any variance in eye color. It is necessary to mate *wa* males with + females and + males with *wa* females because the essence of a sex-linked experiment is based on whether the offspring from *wa* males and + females differ in eye color from the offspring of *wa* females and + males. In broader terms, if A and B represent two identical species differing in the allele being tested, then the outcome of a sex-linked experiment depends on whether the offspring of male A and female B differ in the trait being tested from female A and male B. If there is a difference in eye color, then the traits of the progeny are dependent on the sexes of the parents. Thus, the particular trait under consideration must be on the sex chromosomes. If there is no difference in eye color between the offspring, then the trait is not dependent on the sex of the parents . . . . . .

In order to count and separate male and female flies, it was first necessary to etherize them so they would not fly away during the counting. This was accomplished by pouring them through a funnel whose spout was wrapped with a string soaked in ether. The flies are shaken from their respective vials through a funnel and into a container below the funnel. The ether concentration in the container is sufficient to render the flies unconscious. Using a dissecting scope, we separated the flies according to sex and eye color. We then took vials containing fly agar and yeast and made 10 vials of three + males and five *wa* females, and ten vials of three *wa* males and five + females. To serve as a control, we made ten vials of three + males and five + females and three *wa* males and five *wa* females . . . .

V. Results

The results of each experiment should be presented clearly, without comment, bias, or interpretation. Graphs, tables of data, and figures are often useful here but may not substitute for a verbal summary (at least) of the findings. The most important features of the figures should be pointed out in this section. (Because we feel the illustrative material is so important, we have included a separate section of this chapter illustrating various types of tables, figures, graphs, etc. It follows the discussion of the "literature cited" part of the scientific paper.)

Statistical tests applied to your data are reported in this section although conclusions about your original hypothesis are saved for the discussion section of the paper. Here are some sample results sections:

PREY PREFERENCE AND HUNTING HABITAT
SELECTION IN THE BARN OWL
Stephen J. Fast and Harrison W. Ambrose, III
The American Midland Naturalist 1976, 96 (2):503-507

Results

In Situation 1, both kinds of rodents were present in both habitats. The owl took significantly more rodents from the field habitat (21) than from the woods (7). The comparison between his prey choices (19 *Microtus*, nine *Peromyscus*) borders on significance.

A field-woods hunting habitat choice with *Peromyscus* was offered in Situation 2. The owl took significantly more mice (11) from the field than from the woods (3).

Situation 3 presented the field-woods choice using only *Microtus*. Again, significantly more rodents were taken from the field (12) than from the woods (3).

In Situation 4, which allowed the owl a choice between the two species in the woods only, the number of *Microtus* eaten (12) was twice the number of *Peromyscus* (6). These results, although suggestive, are not statistically significant, probably due to the small sample size.

The owl was permitted to choose between the two prey species again in Situation 5, this time in the field only. Significantly more *Microtus* (10) were eaten than *Peromyscus* (2).

In Situations 1, 4, and 5, where the owl had the opportunity to choose between the two species of prey, a total of 41 *Microtus* and 17 *Peromyscus* were taken. Situations 1, 2, and 3 gave the owl the field-woods choice; a total of 44 rodents were taken from the field and 13 from the woods. These results, when subjected to the X^2 test, show *Microtus* to be significantly the "preferred" prey item, and the field to be the preferred hunting habitat.

REGULATION OF TREHALOSE SYNTHESIS IN THE
SILKMOTH *HYALOPHORA CECROPIA:* THE ROLE OF
MAGNESIUM IN THE FAT BODY
by Arthur M. Jungreis, Peter Jatlow and G.R. Wyatt
Journal of Experimental Zoology, 1974. 187(1):41-45

Results

The concentration of magnesium was determined in the fat bodies of a number of mature feeding cecropia larvae and pupae in diapause and was found to decline by about 50% between these two developmental stages, in rather close correlation with the previously observed change in intracellular trehalose concentration (table 1). During the larval and pupal stages, both magnesium and trehalose undergo marked fluctuations in concentration. In a large series ranging from fourth-instar larvae to pharate adults, the concentrations of magnesium and of trehalose in the hemolymph showed no correlation (fig. 1). When the concentration of trehalose in hemolymph was compared with that of magnesium in the fat body, however, a significant positive correlation was observed (fig. 2). The data in figure 2 are from insects taken one to two days before spinning and one to one and one-half days before the larval-pupal ecdysis (3.5-4 days after applysis), at which times the cells are penetrable to trehalose so that the levels of this sugar in the tissues would be similar to those in the hemolymph.

The lack of correlation of hemolymph magnesium with trehalose, while fat body magnesium is corrected, suggests that movement of magnesium between hemolymph and fat body is restricted. To test this, we injected five fith-instar larvae, taken about one day before spinning, with volumes of 1 M MgCl$_2$ sufficient to raise the concentration of magneiusm in the hemolymph by about 60μEq/ml. After four hours, samples of hemolymph and fat body were taken for analysis; a longer time was not used to avoid complications due to excretion of magnesium (Maddrell, 1971). The results (table 2) show that the elevation of magnesium in the hemolymph was not attended by any rise in the fat body. The uninjected comparison group was taken one day earlier, which may account for its somewhat higher levels of both trehalose and fat body magnesium.

THE *IN VITRO* MAINTENANCE OF THE APICAL ECTODERMAL RIDGE OF THE CHICK EMBRYO WING BUD: AN ASSAY FOR POLARIZING ACTIVITY
by Jeffrey A. MacCabe and Brenda W. Parker
Developmental Biology, 1975. 45:349-357

Results

The *in vitro* behavior of the ectoderm and mesoderm of wing bud responding tissue was examined when cultured with tissue from various regions of the wing, both from outside and within the zone of polarizing activity. The results are described below.

Responding Tissue Cultured Alone

In 73 cases responding tissue was placed in culture and incubated for 24, 36, or 48 hr (Table 1, a). In over half of the 24-hr cultures, the apical ectodermal ridge was not present (Fig. 3a) or was represented only by a remnant at the posterior end of the tissue. In most of the remaining cultures the ridge was present but thin. After 36 hr nearly all of the cultures were without ridge and after 48 hr in culture none had an apical ridge. All of these cultures contained macrophages in the mesenchyme subjacent to the ectoderm (Fig. 3b and c). In most of the cultures the macrophages were scattered along the anteroposterior axis of the responding tissue. In seven out of ten cases with a ridge remaining after 24 hr the ridge was confined to the posterior half of the responding tissue and the cell death to the anterior half. After 36 hr in culture the macrophages were scattered along the entire a-p axis of the tissue. The macrophages in the 48-hr cultures were even more scattered, an occasional one even escaping the responding tissue altogether. The data suggest a slight increase in the number of macrophages with increased time in culture.

133

Responding Tissue Cultured in Contact with Polarizing Tissue

The behavior of the responding tissue *in vitro* was markedly different when cultured in contact with wing tissue from the "zone of polarizing activity" (Table 1, b). In 75 cases polarizing tissue was cultured in contact with either the anterior or posterior end of the responding tissue. In most cases the apical r..uge appeared thick and there were no macrophages apparent after 24, 36, or 48 hr of culture (Figs 4a and b). In the three cases where macrophages were present, they were few. There also . . .

Responding Tissue in Contact with Tissue from Outside the Zone of Polarizing Activity

Two sources of wing bud tissue from regions outside the zone of polarizing activity were placed in contact with responding tissue, the anteroproximal corner and the center of the dorsal surface of the wing bud. In addition, tissue from the flank about midway between the leg and wing buds was also used. When the flank or anterior limb tissue was placed in contact with responding tissue the ectodermal ridge of the responding tissue . . . .

PHEROMONES IN TRIBOLIUM
by Cary Miller
(student — University of Illinois)

Results

The results of the four experimental groups and their respective controls are as follows:

In Table I, the mean and standard deviation both refer to the number of beetles counted within the circle of the test box.

Type of Tribolium in box	Type of Tribolium in cup	Mean	Standard Deviation
McGill McGill	McGill none	67 22.2	4.64 3.56
McGill McGill	Ebony none	61.8 24.4	5.4 4.55
Ebony Ebony	Ebony none	64.6 24.4	4.88 4.83
Ebony Ebony	McGill none	64.6 25.6	7.02 4.16

Table I. The experimental setup and the means and standard deviations of the number of beetles counted in the response circle of each test situation.

In all cases the F-test showed the variances not to be significantly different. Thus, we used the formula for finding t when the variances are equal. Also, for all of our trials, the degrees of freedom is equal to 8. The t calculated for the McGill box with McGill cup test was 6.40. The corresponding critical was 2.306, thus we rejected our null hypothesis. Since there were more beetles in the circle in the experimental group than in the control group, we chose the hypothesis that McGill are attracted to each other and consequently have pheromones. In the McGill box-Ebony cup test, the calculated value for t was 6.34, considerably greater than the critical value of 2.306. As a result, we rejected our null hypothesis. Again, since there were more beetles in the circle in the experimental group than in the control group, the hypothesis that said McGill were attracted to Ebony was chosen. Furthermore, we attributed this attraction to pheromones.

In the Ebony-McGill cup test . . . .

COMPETITIVE EXCLUSION
by Sue Kirk and Nancy Sagmeister
(students — University of Illinois)

Results

Our results showed that in 27 *E. coli* and *Pseudomonas* crosses in which the plates had been streaked with equal amounts of bacteria, 21 plates were dominated by *Pseudomonas* and 6 were dominated by *E. coli*.

VI. Discussion

In this section you evaluate the meaning of your results in terms of the original question or hypothesis and point out their biological significance. If the results are unexpected or contradictory, you should attempt to explain why and possibly point out avenues of further research. The discussion should describe the significance of your experiment in terms of other work without trying to review the entire field. Pertinent, related literature should be cited, however, for purposes of comparison.

DISTASTEFULNESS AS A DEFENSE MECHANISM
IN *APLYSIA BRASILIANA* (MOLLUSCA: GASTROPODA)

by H.W. Ambrose III, R.P. Givens, R. Chen, and K.P. Ambrose

(submitted for publication)

Discussion

The results indicate the presence of a presumably distasteful substance which causes gulls to reject significantly more pieces of *Aplysia brasiliana* than fish. None of our observations indicated that *Aplysia brasiliana* was toxic to the gulls, although occasionally a gull which appeared to have swallowed a test piece without tasting it first was seen shaking its head with its mouth open.

As would be expected in a mechanism of defense against predation, the outer, most vulnerable, portions of *Aplysia brasiliana* were the most distasteful. This supports Thompson's suggestion (1960a and b) that the defensive secretions are present in greatest abundance in the areas of skin which would first be encountered by a predator.

It is interesting that any pieces of *Aplysia brasiliana* were eaten at all since few researchers have been able to feed *Aplysia* to other species. Watson (1973) had little luck force-feeding midgut gland to mice and Ambrose (unpublished data) was unable to feed fresh *Aplysia* to pinfish. However, all of our test pieces were between one and two hour's "dead" and Thompson's data (1960a) indicate that dead opisthobranchs lose some of their distasteful properties.

It is perhaps of greater interest that gulls did not discriminate between fresh fish and pieces that had been soaked in the blood, mucus, and ink — a mixture one might expect to be highly distasteful. *Aplysia* ink has traditionally been assumed to play a defensive role (Hyman, 1967; MacMunn, 1895), although it is no longer held that this rather sluggish species makes an escape behind the screen of a cloud of ink. It has been shown that even in the smallest tidepool, the ink does not actually obscure the animal (Kupfermann and Carew, 1974). Other theories about the ink can be found in the literature. Chapman and Fox (1969) suggest that it is merely a waste product of algal metabolism and Tobach et al. (1965) suggest that it communicates information about reproductive states because solitary animals ink less often than those in a group. It has been our experience that *A. brasiliana* invariably ink when first scooped out of the water in a net and often ink when undergoing a physical examination unless handled very gently. We would like to offer another theory about the ink which does credit it with a defensive role. When Thompson discusses the defensive adaptations of opisthobranchs (1960b), he points out that his data are not consistent with the usual assumption that brightly colored species are normally toxic or distasteful and that cryptically colored species are usually edible. He points out that the fairly cryptic *Aplysia punctata* is as distasteful as the more conspicuously colored opisthobranchs. We would like to suggest that the purple ink of *Aplysia brasiliana* serves as its warning coloration and conclude that this species escapes predation by being a highly distasteful animal and by warning potential predators of this fact with a cloud of purple ink.

INSENSITIVITY OF LEPIDOPTERAN TISSUES TO OUABAIN: PHYSIOLOGICAL MECHANISMS FOR PROTECTION FROM CARDIAC GLYCOSIDES

by Gerald L. Vaughan and Arthur M. Jungreis
Journal of Insect Physiology, 1977. 23:585-589

Discussion

The Na^+-K^+-ATPase is generally responsible for alkali cation transport in animal cells (Skou, 1957) and is inhibited by ouabain and similar cardiac glycosides (Glynn, 1957). The Lepidoptera used in this study and perhaps similar phytophagous insects are unusual in several respects, among which is the absence of measureable Na^+-K^+-ATPase in tissues other than those rich in neuronal material (Vaughan and Jungreis, 1976; Jungreis and Vaughan, 1976, 1977). In addition to having epithelia which appear to lack Na^+-K^+-ATPase and concomitant cardiac glycoside sensitivity, these insects differ from other animals in at least one other important respect — high levels of K^+ and low levels of Na^+ in blood (Florkin and Jeuniaux, 1974; Jungreis, 1977). The location of Na^+-K^+-ATPase in neuronal tissue, however, and the K^+ and Na^+ content of tissues from the head of *D. plexippus* are evidence that these insects maintain the same mechanisms for sustaining action potentials in the central nervous system as do other animals whose blood is high in Na^+ and low in K^+ (Treherne, 1967, 1976).

The physiological character of the insects in this study and the information reported here suggest a convincing mechanism for the insensitivity of phytophagous Lepidoptera to cardiac glycosides. Considering the refractory nature of *D. plexippus* to cardiac glycosides, it is not unexpected (Jungreis and Vaughan, 1977) that Na^+-K^+-ATPase from this organism would have a low affinity for ouabain and thus the low sensitivity noted in Fig. 4. More interesting is the lack of observable *in vivo* perturbation in *M. sexta*, which possesses Na^+-K^+-ATPase highly sensitive to ouabain *in vitro* (Fig. 4). We suggest that following introduction of cardiac glycoside but prior to storage, degradation or excretion (Duffey and Scudder, 1972; Brower and Glazier, 1975), high haemolymph K^+ lessens or totally blocks (Fig. 5) the effect of the glycoside on nervous tissue, possibly the only tissue containing Na^+-K^+-ATPase. During this period of K^+ protection, glycosides are either rapidly excreted or stored in various tissues. Based on this model and the suggestion that the neuronal sheath (Treherne, 1967, 1976) affords additional protection, these insects would not need other special enzymatic adaptations to cope with the problems of cardiac glycoside ingestion.

The slight but significant sensitivity to the Na^+-K^+-ATPase in *D. plexippus* requires additional interpretation. It offers an obvious explanation for the observations (Urquart, 1961; Brower and Glazier, 1975) that Monarchs with the highest levels of cardiac glycosides are smaller and less successful. The total amounts of glycoside they contain (Brower and Glazier, 1975) are comparable to those used in this study. The so-called physiological burden (Brower and Glazier, 1975) of cardiac glycoside ingestion in the Monarch butterfly had been considered to be a function of the cost of metabolizing and storing the cardiac glycoside with little regard for the possibility that the insect itself may possess cardiac glycoside-sensitive moieties. A reassessment of the true situation must accommodate the toxic effects of very high levels of cardiac glycoside on neuronal Na^+-K^+-ATPase as well as the metabolic cost of sequestration.

RARE MALE SELECTION IN *DROSOPHILA*
by Susan Hemingway
(student – University of Illinois)

Discussion
In collecting and analyzing the data, many assumptions were made:
1) Female mating behavior does not change in differing social environments, that is, in an environment containing a greater number of males than females.
2) All the females were equally receptive to mating at the time of exposure to males.
3) No males were sterile.

4) Females maintain a consistency in their selection of mates.

The results show only that the female mates with the male strain that is least frequent in the population. If the frequency of the different male strains are equal in the population, then the females mate randomly. There is an equal probability of mating with each male.

The question still remains whether the female is actually *selecting* a particular male or whether some other process is involved. It may be that a male becomes more aggressive when he is outnumbered by individuals of another strain and therefore succeeds in mating with the female. Perhaps the female undergoes a physiological change that prevents her from mating or bearing the young of the more abundant male type. There are a vast number of possibilities that could account for the rare male selection behavior in *Drosophila* and a great deal of research is needed before the actual cause of this behavior can be narrowed down and supported.

There is no reason to doubt that the ebony-body strain of *Drosophila* exhibits this behavior. Rare male selection behavior has been observed among strains of *Drosophila* in which the males have been phenotypically different from one another as well as when the males differed only in their geographical origin. By testing different strains, it may be possible to discover whether the behavior is representative of only certain strains, and if so, is it the male's strain, the female's strain, or both that determine whether the behavior will take place.

A COMPARISON OF THE DIVERSITY OF ORGANISMS IN TWO MAN-MADE PONDS
by Dave Dohse
(student — University of Illinois)

Discussion
An age difference of approximately five years between the I-74 and I-72 ponds was shown not to "cause" a significant difference in the diversities of the two ponds. One explanation of this equality in diversity could be that both ponds may be old enough to have reached an ecological or community equilibrium despite their age difference. Certainly the five year age difference between the two ponds seems to be significant, but even the youngest of the two ponds (I-72) has been in existence for over five years (1970-1976). This explanation seems to fit the findings quite well. However, this does not offer any answer as to how soon the equilibrium might be established. A worthwhile answer would only be possible

after a more extended investigation which followed the diversity of a pond over the course of several years.

Although there were many chemical factors involved which could have caused significant differences in the results, a concurrent investigation concerning the chemical differences in the water of two similar, man-made ponds hopes to show these chemical differences as negligible. Whether a significant chemical difference is shown or not, however, it seems as though any difference in diversity for the two ponds is not related to their ages. Better control over chemical, meteorological, and sampling factors would certainly aid in determining the exact role played by the age of a pond in terms of its diversity.

The somewhat greater numbers of Eucrustaceans (Brachiopoda, Copepoda, Malacostraca) observed in this investigation could very well be due to the increase in population numbers of these organisms during spring and fall, as has been shown by Whiteside (1972). However, support for Whiteside would only be possible if data for all seasons were available.

The absence of any other studies dealing with the relation between age and organism diversity in the *substratum* of freshwater ponds is certainly cause for further investigations of this kind.

RESISTANCE TO INSECTICIDE IN *DROSOPHILA MELANOGASTER*

by Barb Pomeranke, Kathy Brockett, and Ron Fitzanko
(students — University of Illinois)

Discussion

In the evolution investigation, the H_1 was not rejected. This would indicate that resistance to insecticide can be selected for in fruit flies.

In the genetics investigation, only the H_3 was not rejected indicating that resistance is inherited in fruit flies. It seems also that resistance to insecticide is a dominant trait and that the original virgin females could have been homozygous for the dominant trait. If non-resistance to insecticide is a recessive trait, the non-resistant males that were crossed with the resistant females must have been homozygous recessive. The cross would then result in 100% heterozygous offspring which would prove resistant to the insecticide. This appears to be the case since 38/40 (95%) of the offspring from this cross survived on the agar that was injected with insecticide. Of course, to verify this assumption about genotypes, additional crosses would have to be made. The fact that the resistant offspring survived equally well on injected agar (vials h and i) as on normal fly media (vials l and m), only indicates that the flies have expanded their "evolutionary niche."

VII. Literature Cited

All published work mentioned in your paper must be listed in this section. You will have saved hours of pain and suffering if you are now working from a neat stack of 3x5 cards, each containing all the bibliographic information needed (see p. 102) for a proper citation. List your citations in order of the last name of the first author. In cases of more than two authors, you may use an "et al." in the text of your paper (Jones et al., 1972), but in the literature cited section, all authors must be named. Here are some sample citations to books and journal articles. Journal titles may be abbreviated according to standard practices (see the *CBE Style Manual* when in doubt).

Literature Cited

Ambrose, H.W., III. 1972. Effect of habitat familiarity and toe clipping on rate of owl predation in *Microtus pennsylvanicus*. J. Mammal. 53:909-912.

Brown, L.E. 1956. Field experiments on the activity of the small mammals, *Apodemus*, *Clethrionomys* and *Microtus*. Proc. Zool. Soc. London 126:549-564.

Council of Biology Editors, Committee on Form and Style. 1972. CBE style manual. 3rd edition. American Institute of Biological Sciences, Washington, D.C.

Eadie, W.R. 1954. Animal control in field, farm and forest. The MacMillan Co., New York.

Hairston, N.G., F.E. Smith, and L.B. Slobodkin. 1960. Community structure, population control and competition. Amer. Nat. 94: 421-425.

Link, G.K.K. 1928. Bacteria in relation to plant diseases. Pages 590-606 in E.O. Jordan and I.S. Falk, eds. The newer knowledge of bacteriology and immunology. University of Chicago Press, Chicago, Ill.

Smith, C.R. and M.E. Richmond. 1972. Factors influencing pellet egestion and gastric pH in the barn owl. Wilson Bull. 84: 197-186.

VIII. Illustrative Material

Frequently a scientific article must be augmented with maps indicating the location of a study, diagrams of a study area or apparatus, figures showing a graphical or diagrammatic representation of the data, or tables of actual results. In all cases, the more complete the labeling, the more effective the presentation of information. Some samples of illustrative material are given below.

MAPS AND DIAGRAMS

This map is taken from Arthur C. Echternacht, 1977. A new species of lizard of the genus *Ameiva* (Teiidae) from the Pacific Lowlands of Columbia. *Copeia* 1977(1):1-7.

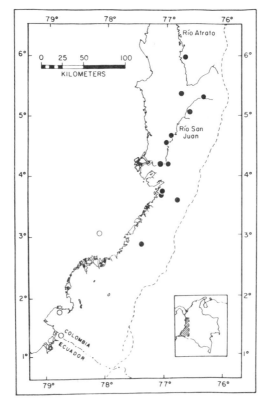

Locality records of *Ameiva anomala* (solid symbols) and Colombian records of *Ameiva bridgesi* (hollow symbols). The off-shore locality for *bridgesi* is Isla Gorgona, Depto. Cauca. The dashed line represents the approximate position of the western cordillera of the Andes.

This diagrammatic illustration of potential relationships is taken from: S.E. Riechert and C.R. Tracy, 1975. Thermal balance and prey availability: bases for a model relating web-site characteristics to spider reproductive success. *Ecology* 56 (2):265-284.

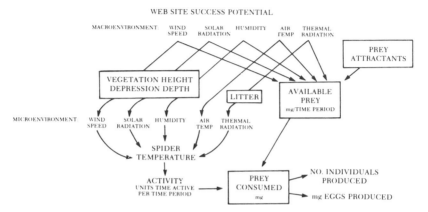

Hypothesized relationship between web-site character and individual reproductive success. Solid lines indicate direct parameter effects on quantities or functions.

This example of a diagram which illustrates a biological process is taken from: G.L. Whitson, 1965. The effects of actinomycin D and ribonuclease on oral regeneration in *Stentor coeruleus. Journal of Experimental Zoology* 160 (2):207-214.

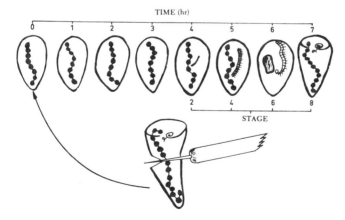

A diagram of oral regeneration in bisected stentors. Regeneration occurs in 6-8 hours. The stages shown are those given by Tartar ('61). The beaded macronucleus coalesces and renodulates late during oral regeneration.

GRAPHS

Graphs can give a quick, visual illustration of significant trends in experimental data. This first graph is taken from: Roland M. Bagby, 1974. Time course of isotonic contraction in single cells and muscle strips from *Bufo marinus* stomach. *American Journal of Physiology* 227(4): 789-793.

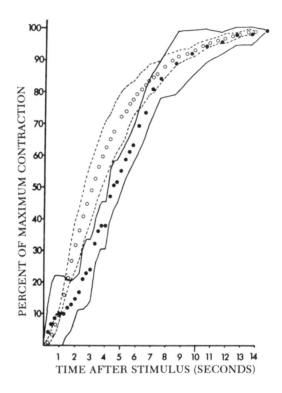

Comparison of mean time course of samples of cells (open circles) with mean time course of muscle strips$_C$ (solid circles). Original data are rescaled so that individual responses use their own final lengths as 100% maximum contraction. Dashed lines and solid lines outline 95% confidence limits for cells and strips$_C$, respectively.

144

This next graph is taken from: A.M. Jungreis, P. Jatlow, and G.R. Wyatt, 1974. Regulation of Trehalose synthesis in the silkmoth *Hylophora cecropia:* the role of magnesium in the fat body. *Journal of Experimental Zoology* 187(1):41-45.

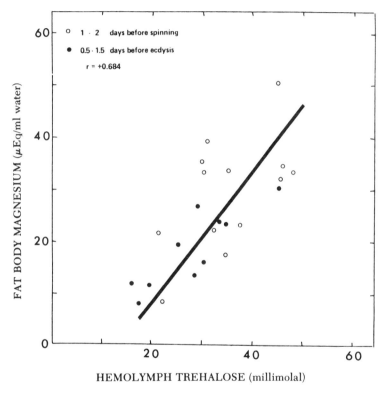

Linear correlation analysis between the concentrations of trehalose in hemolymph and magnesium in fat body in larval and pharate pupal stages of development.

This graph was taken from: G.L. Whitson, J.G. Green, A.A. Francis, and D.D. Willis, 1967. Cyclic changes in the alcohol-soluble carbohydrates in synchronized *Tetrahymena. Journal of Cellular Physiology* 70(2):169-177.

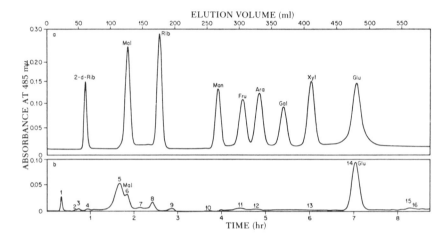

Chromatograms of samples of (a) 9-components of an 18-component standard containing 0.5 micromoles of each sugar and (b) 1 ml extract of synchronized cells obtained 75 minutes after EHS. Peaks are numbered arbitrarily for identification. Note that peak 6 (maltose) and peak 14 (glucose) are two of the major peaks identifiable by parallel chromatography with the standard (a).

The next two graphs are taken from: Susan E. Riechert and C. Richard Tracy. Thermal balance and prey availability: bases for a model relating web-site characteristics to spider reproductive success. *Ecology*. 1975. 56 (2):265-284.

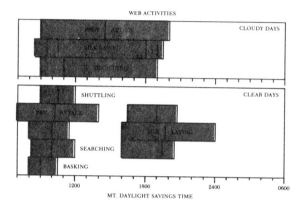

Periodicity of various web activities for clear and cloudy days. Bars represent time periods in which 90% of each of the designated activities were observed. Vertical lines represent the medians of these activities.

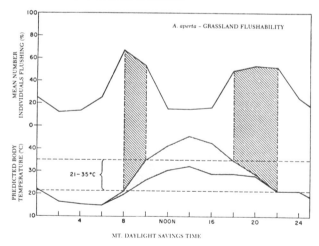

Graph of percent of spiders active on the sheet with time of day on the mixed-grassland study area in midsummer (July and August) imposed on predicted spider temperature under these conditions and assuming a web-over-litter substrate. Barred area under flushability curve represents time periods during which over 50% of the individuals were active. Stippled area represents range of spider temperatures, exact temperature dependent upon amount of exposure to solar radiation. Upper boundary of predicted temperature curve signifies spider temperature if in full sunlight. Lower boundary signifies spider temperature if in full shade. Area enclosed by dashed lines represents body temperature range within which over 50% of the spiders are active.

147

The last two graphs are from: H.W. Ambrose, III, 1973. An experimental study of some factors affecting the spatial and temporal activity of *Microtus pennsylvanicus. Journal of Mammalogy* 54 (1):79-110.

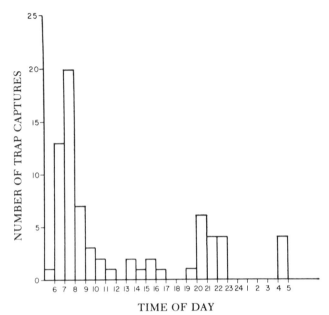

TIME OF DAY

Relationship of trap captures to time of day for all study voles.

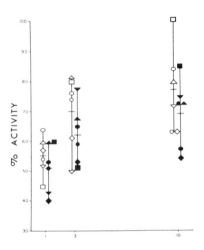

MICE PER PEN

Mean percentage activity of individual male (closed symbols) and female (open symbols) *Microtus* at three population densities. Individuals are indicated by differently shaped symbols.

148

TABLES

A great deal of detailed data can be presented in tabular form and merely summarized verbally in the text of an article. Some sample tables follow.

The first table is taken from: A.M. Jungreis and G.R. Wyatt, 1972. Sugar release and penetration in insect fat body: relations to regulation of haemolymph trehalose in developing stages of *Hyalophora cecropia*. *Biological Bulletin* 143:367-391.

COMPOSITION OF HAEMOLYMPH OF *Hyalophora cecropia* REARED ON STANDARD OR SPECIAL ARTIFICIAL DIETS

Stage	m-equiv/l.					Trehalose (mM)	Osmotic pressure (milli-osmolal)
	K	Na	Mg	Ca	Cl		
Larvae							
Mid-fourth instar	28	20	56	9·2	33	35	300
Early fifth instar	35	11	66	12	33	35	307
Early fifth instar, special diet*	44	16	66	9·3	—	33	360
Mid-fifth instar	32	12	69	10	33	44	255
Mid-fifth instar, special diet*	35	9·0	76	6·5	46	41	313
Mature fifth instar (1)†	20	—	40	10	36	44	266
Mature fifth instar (2)	33	5·7	65	8·2	24	45	297
Larval-pupal transformation							
First day of spinning (1)	21	—	40	10	24	35	280
First day of spinning (2)	24	1·4	37	—	22	54	289
Second day of spinning (1)	22	—	45	11	24	31	376
Second day of spinning (2)	30	1·3	54	11	—	44	279
Second day after apolysis (1)‡	33	—	63	12	22	—	311
Second day after apolysis (2)	36	1·3	57	12	29	44	—
Third day after apolysis (1)	31	—	48	12	22	23	287
Third day after apolysis (2)	48	2·3	53	12	22	36	—
Pupae							
Day of ecdysis (1)	39	—	37	11	20	28	373
Day of ecdysis (2)	49	1·6	51	10	23	28	354
+ 1 week 25°C	35	<2	36	11	18	17	343
+ 2 weeks	31	<2	36	10	18	16	480
+ 4 weeks	39	<2	35	11	22	17	584
+ 8 weeks	37	<2	27	11	19	8	>1000§
+ 8 weeks 25°C and 8 weeks 6°C	36	—	46	9·3	—	20	—
Diet, standard	57	19	15	56	35	—	>325
Diet, special	82	19	65	56	110	—	>400

*The standard diet was supplemented with 25 mM $MgCl_2$ and 25 mM KCl.

†(1)and (2) signify samples from the first and second experimental series, respectively (see Materials and Methods).

‡Apolysis occurred on the third day after the beginning of spinning.

§Values in apparent excess of 1000 milliosmolal represent samples which failed to freeze under the supercooling conditions used, because of their high content of glycerol.

This next table is taken from: Jeffrey A. McCabe and Brenda W. Parker, 1976. Evidence for a gradient of a morphogenetic substance in the developing limb. *Developmental Biology* 54:297-303.

MORPHOGENETIC ACTIVITY IN NORMAL AND EXPERIMENTAL 5-DAY WINGS

Site tested for activity	Number of cultures	Responding tissue after 24 hr of culture			Average number of macro-phages per culture
		Ectodermal ridge			
		Thick	Thin	None	
a. Anterior border	35	2	11	22	44.5
b. Middle of limb	37	25	10	2	17.9
c. Posterior border	33	30	3	0	7.3
d. Middle with posterior Mylar barrier	32	4	12	16	49.9
e. Posterior border with posterior Mylar barrier	40	29	11	0	4.0
f. Middle with anterior Mylar barrier	32	11	13	8	16.4
g. Middle with posterior filter barrier	32	11	14	7	19.4
h. Middle, posterior one-third excised	37	4	16	17	45.0
i. Middle, anterior one-third excised	31	3	11	17	33.7
j. Middle, apical one-third excised	33	5	13	15	24.5

This table is taken from: S.J. Fast and H.W. Ambrose, III. Prey preference and hunting habitat selection in the barn owl. *The American Midland Naturalist.* 1976. 92 (2):503-507.

The number of rodents eaten of each species in five different combinations of species and habitat types is shown. To the right are the X^2 values for each comparison. * = significance at the .05 level, 1 df. wM and wP = *Microtus* and *Peromyscus* familiarized with the woods habitat. fM and fP = *Microtus* and *Peromyscus* familiarized with the field habitat

	Number of Rodents Eaten				
	Field Habitat		Woods Habitat		
	fM	fP	wM	wP	
SITUATION 1					wM + wP vs. fM + fP
Replicate a	5	2	2	0	X^2 = 7.00*
Replicate b	4	2	0	1	wM + fM vs. wP + fP
Replicate c	5	3	3	1	X^2 = 3.57 N.S.
Total	14	7	5	2	
SITUATION 2					
Replicate a	—	6	-	1	
Replicate b	—	5	-	2	
Total	—	11	-	3	X^2 = 4.57*
SITUATION 3					
Replicate a	5	—	0	-	
Replicate b	7	—	3	-	
Total	12	—	3	-	X^2 = 5.40*
SITUATION 4					
Replicate a	—	—	9	3	
Replicate b	—	—	3	3	
Total	—	—	12	6	X^2 = 2.0 N.S.
SITUATION 5	10	2	-	-	
Total	10	2	-	-	X^2 = 5.33*

151

Our final example of a table is from H.W. Ambrose, III, 1972. The effect of habitat familiarity and toe-clipping on rate of owl predation in *Microtus pennsylvanicus*. *Journal of Mammalogy* 53 (4):909-912.

THE EXPERIMENTAL DESIGN AND RESULTS

Exp. Test Unit	Number voles available	Predator hours per vole eaten	Toe-clipped voles eaten	Non-clipped voles eaten
Mice Familiar with environment				
1	10	38	4	3
2	10	10	2	3
3	7	24	0	3
4	10	43	4	1
5	8	24	4	1
Mice unfamiliar with environment				
1	10	5	3	2
2	10	4	3	4
3	10	14	2	3
4	10	10	2	3
5	10	6	5	3

Chapter Eleven
Problems and Pitfalls
In Writing a Scientific
Paper

I n this chapter we will touch upon a few of the more common technical difficulties associated with writing the scientific paper. For more detailed help we strongly recommend the *CBE Style Manual* and another Council of Biology Editors' publication, *Scientific Writing for Graduate Students*, edited by F. Peter Woodford.

I. Citing other work.

Factual information taken from published sources should be documented whether or not you use an actual quotation. Footnotes are seldom used in scientific writing; generally, reference in the text is made only to the author's name and date of publication — the full bibliographic information appears only in the Literature Cited section. Both the name and date can go inside parentheses if the name is not actually part of your sentence. For example:

> *A large part of the natural diet of barn owls consists of* Microtus *(Wallace, 1948; Craighead and Craighead, 1956)* . . .

If the author's name is intended to be part of the text, only the date goes in the parentheses:

> *Metzgar (1967) has shown that* . . .

And when the date too is important to your text, you can omit the parentheses.

> *As early as 1967, Metzgar showed that* . . .

If there are more than two authors, the citation in the text can be condensed to an "et al." form; however, the full citation must appear in the Literature Cited.

. . . theory of population regulation (Hairston et al., 1960).
Hairston et al. (1960) have suggested that . . .

II. Polishing your paper.

A direct, concise, precise style can be achieved only through practice. Always allow time for two drafts of a paper. When you have put all your ideas down, read back through them very slowly and carefully, considering each word and sweeping out extraneous verbiage and faulty or misleading grammatical constructions.

In this chapter we will give some typical examples of sloppy, unpolished writing taken again from student papers. As you read the examples, try to think of how to improve them before you read the suggested corrections. This exercise will help you spot similar difficulties in your own papers.

Wordy, awkward constructions.

It is thought that possibly the reason why Hawaii does not fit the area-species curve of the islands around it is because of its recentness.

(It is possible that Hawaii does not fit the area-species curve of the islands around it because of its geological recentness.)

However, when dealing with small microorganisms, identifying them apart from each other is quite difficult.

(However, microorganisms are difficult to identify.)

The results may be restated to say that the directed movement response of female Acheta domesticus *to a tone versus silence, a cricket song versus silence, and a cricket song versus a tone, is not significant in terms of which one the insects will move toward.*

(Restating the results, there was no significantly directed movement in the female *Acheta domesticus* in response to a choice between a tone versus silence, a cricket song versus silence, and a cricket song versus a tone.)

By our research we are attempting to test whether the minority effect is valid.

(We are attempting to test the validity of the minority effect.)

How species may compete with each other has been based on models of how crowded animals of the same kind may so compete among themselves that their population growths are curbed and their crowds are controlled.

(Possibly this means: Mathematical models have been constructed demonstrating that the growth rate and density of similar species populations are restricted by competition.)

In the case of the other three pairs of resident and intruder males, the result of putting an intruder male in the cage of a resident male produced different results.

(Putting an intruder male in the cage of a resident male produced different results in the other three trials.)

The question we are asking is: Using three different habitats, what is the diversity and abundance of animals within each habitat, and how do they differ?

(How do the diversity and abundance of animals differ among the three different habitats?)

155

It appears by looking at Graph A that . . .

(Graph A shows that . . .)

Often wordy sentences can be improved by removing the passive voice.

In an investigation by Hutchinson of the animals of an intertidal rocky shore, it was noticed that the adults of two species of barnacles occupied two . . .

(Hutchinson noticed that two species of barnacles, on an intertidal rocky shore . . .)

Dangling participles and other constructions in which a sentence somehow gets the **wrong subject** can be embarrassing.

Using sterile technique, the bacteria . . .

After injecting the hormone, the frog . . .

By studying competition, ecology can better understand . . .

Numerous experiments within the last ten years have reported that the mutagenic, carcinogenic, or chromosome damaging effects of . . .

(Here the experiment has to do its own reporting)

Try to **use verbs instead of abstract nouns.** Expressions such as "was accomplished," "was performed," and "were attained" should be warning signals. These phrases are pompous.

The calculation was performed (we calculated . . .)

The measurement was attained (we measured . . .)

Often a paper lacks precision because a word or phrase has not been used properly.

The results of the experiment demonstrate that fish can not be taught certain behavior, conditioned, and thus fish learn behavior pattern by instinct.

(Here the student makes a rather grandiose conclusion based on his failure to condition a fish and jumps unwittingly into the foggy area of learning versus instinct by carelessly implying that things can be "learned by instinct.")

When two species are competing for the same space, usually one will overpopulate the other.

(Presumably the student meant simply that one species might outnumber the other but permitting "overpopulate" to have a direct object suggests that possibly he means "eliminate.")

Out of the 45 plates we streaked, Pseudomonas *dominated the bacterial cultures in 33 of the plates.*

(The sinister political implications of "dominating a culture" are unintentional. Presumably "predominated in the cultures" is implied.)

... the males began with aggressive stridulations towards the females.

(Can a cricket stridulate *towards* another cricket? Probably not.)

Supposedly smaller islands hold smaller populations and are therefore subject to faster extinction.

(Does this mean that the islands are supposedly smaller but we are not sure? Or that smaller islands supposedly hold smaller . . .)

Fish have commonly been known to associate certain smells, colors, or sounds to mean different things.

(They may well associate these things *with* other things but that is all.)

Avoid incomplete sentences.

On our first day we obtained two stocks. One of white apricot mutant fruit flies and the other of red-eyed wild-type fruit flies.

(On our first day we obtained two stocks — one of white . . .

On our first day we obtained two stocks. One was white . . .)

Beware of ambiguous antecedents.

Onion bulbs were partially submerged in water with the root end just below the surface. Three toothpicks were affixed to the onions to hold them in place. These were grown for three days.

(Could the toothpicks have grown any in that short time?)

If the number of protozoans on the islands have been reduced during dispersal to remote places, then they have successfully attained the distance effect.

(Have the protozoans or the remote places attained the distance effect?)

It has been found that the group diversities of these organisms differ with the habitats, which can be attributed to the influence of different physical characteristics in the habitats.

(It would be clearer to say that the fact that the group diversities differ can be attributed . . . and avoid the confusion over "habitats which . . .")

Needless to say, **it is incorrect to change the form of the verb mid-sentence** . . .

We observed which flies were male and which were female by first knocking them out with ether and then, using a magnifying glass, we looked at the back of the flies.

(first knocking, then looking)

. . . or to give a **plural subject** a **singular verb** form.

Traits of immunity enables the species to continue because . . .

(Traits enable . . .)

And as a final kindness to your readers, **break up noun clusters and stacked modifiers.**

One cricket-call cassette tape recording was obtained on a . . .

(We obtained one cassette tape recording of a cricket call . . .)

As a final suggestion, just before you sit down to hone and pare your writings, consider this poem.

POEM

If you've got a thought that's happy,
   Boil it down.
Make it short, and crisp, and snappy —
   Boil it down.
When your brain its coin has minted,
Down the page your pen has sprinted,
If you want your effort printed,
   Boil it down.
   — Anonymous

159

Chapter Twelve
Illustrating Research Papers
by
Wynne Brown
Brownline Graphics
Knoxville, Tennessee 37938

Many people (and not just students) find "bare" statistics frightening and difficult to understand. One of the ways to make your data more approachable is to present it pictorially. Not only will your material be less intimidating, but it will also be easier to understand if accompanied by a carefully planned figure.

In Chapter Two you learned that "the kind of data you collect will dictate the way the data can be analyzed." The same is true of illustrations: the kind of data dictate the kind of illustrations. Three types of figures are used most frequently: maps, graphs, and charts. Another way to show your data is in the form of a typed table. These are not as effective as illustrations but occasionally some data won't fit into any other sort of framework.

The first question to ask yourself is: how is this illustration to be used? Will it be in a 8 1/2 x 11" format, as in a report? Or will it be published in a scientific journal? If you are writing a thesis or dissertation, what margin widths are required by your particular university? Will you need to have slides made for a presentation? If so, it is easiest on the viewer if all your slides are either vertical or horizontal.

You will save yourself a great deal of time if you answer these questions **before** drawing your rough draft. Then use the appropriate format.

The Rough Draft

Many students are fortunate enough to have access to a computer. There is a great deal of software available to use for charts, graphs and illustrations. Although the market changes frequently, it is time well invested to find a software program that you can use and to become comfortable with it. Some examples of computer graphics are shown in Figure 12.1.

Whether you're doing your figures by hand or on computer, you will still need to plan them carefully. Many students make the mistake of rushing through the rough draft, instead of realizing that it is the most important stage. The rough draft is the real content of your illustration while the final draft is mostly cosmetics.

Generally speaking, if your rough draft fits comfortably on an 8 1/2 x 11" sheet when all the lettering is no less than 4 mm tall, it will be suitable for all uses: reports, slides, publication and poster sessions. The letter height needs to be at least 4 mm since smaller labels will become illegible if reduced for publication later.

When designing any kind of illustration, ask yourself, "What am I trying to show?" and "What is the most effective way to show it?" Always remember to **keep it simple**. If you are making a map of the study site where you counted trout in different streams, don't include all the hiking trails, contours, county boundaries and survey markers. Eliminate everything that will detract from the message you want to convey. Also eliminate all unnecessary text. You worked hard to collect your data: allow them to speak for themselves.

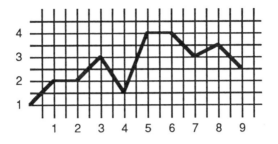

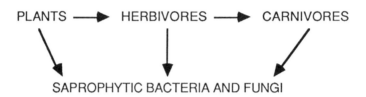

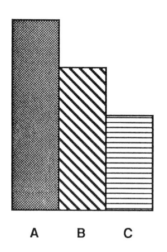

Figure 12.1. Examples of computer graphics.

Maps

A map is useful when you want to show the distribution of data over a geographic area (see any field guide for examples of species localities) or if you wish to show a diagram of your study site (Figure 12.2). Be sure to indicate scale, either with an inset map, geographic coordinates, or with a kilometer/mile scale. The simplest way to show north is to use a capital N inside a hollow arrow. It is also important to indicate where the site is: a detailed map of John Hands Pass is much more meaningful if you include an inset map showing its location in southeastern Arizona. If the site you're discussing is near water or is an island, the reader will have an easier time if you use a shading film (Zipatone and Letraset both make one with wavy lines) or other means to make the water look different from land. If the amount of water is small, you can also fill it in with solid black (Figure 12.3). The names of bodies of water hould be in italics, wherever possible.

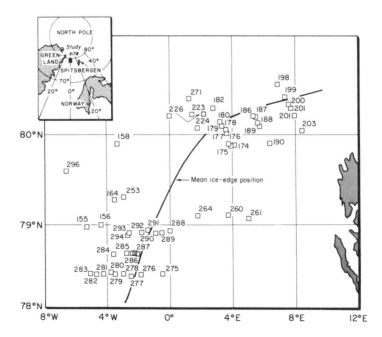

Figure 12.2. Example of a map of a study site. (From Smith, Sharon L., Walker O. Smith, Lewis A. Codispoti, and David L. Wilson. Biological observations in the marginal ice zone of the East Greenland Sea. Journal of Marine Research, vol. 43, (pp. 693-717).

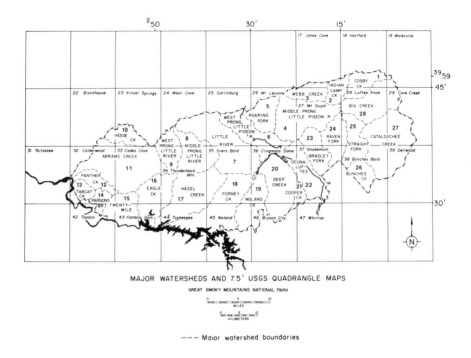

Figure 12.3. Map of Smokies showing lake as black.

A cautionary note: If you are using hollow circles as data points on your map, do not put a town name immediately to the right or left of the data point. The town of Alice Springs has become Alice Springso in Figure 12.4.

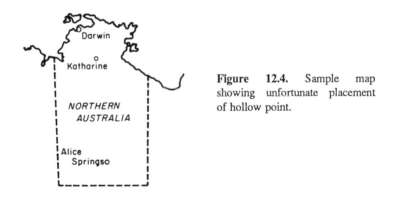

Figure 12.4. Sample map showing unfortunate placement of hollow point.

164

Graphs

There are three types of graphs: bar (also called histograms), line (fever charts) and scatter graphs. Bar graphs are shown in Chapter Three and are usually rectangles representing categories of discrete data. The rectangles can be contiguous as in Figure 3.2 or you can allow space between them (Figure 12.5) for a more graceful figure.

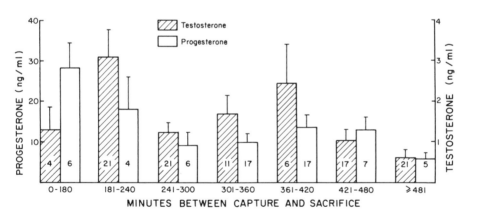

Figure 12.5. A bar graph. (From Wade, J.K. and A.C. Echternacht. Annual hormonal changes in the lizard *Anolis carolinensis*. Gen. Comp. Endocrin. [Accepted for publication.])

Scatter and line graphs are most commonly used and can be drawn together or separately. Data that are continous are usually illustrated with a line graph with all data points indicated. Two line graphs appear on page 42. Discrete data can be illustrated with scatter graphs, such as the ones used in the beginning of Chapter Six. The two can also be combined as in a case where you want to show standard means or a linear correlation (Figure 12.6).

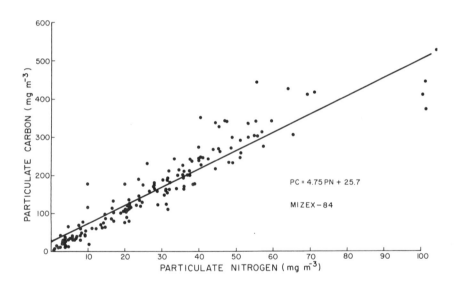

Figure 12.6. Combined scatter and line graphs. (From Smith, W.O., M. Baumann, D.L. Wilson and B. Aletsee. Productivity and biomass in the marginal ice zone of the Fram Strait. Journal of Geophysical Research [In press])

Again, keep your graphs as simple and as easy to read as possible. Here are some points to remember:

1. Use graph paper for your rough draft. It comes in many different grid sizes; find a sheet that fits the size and shape of your graph.
2. Combining two or more figures into one may give greater impact (Figure 12.7).
3. Use capital letters on the abscissa and ordinate except when abbreviating measurements ("min" for minutes, "hr" for hours, etc.). Use lower case letters for any text inside the graph.
4. Select your symbols with care: for example, solid hexagons when reduced are very similar to solid circles. Several companies make inexpensive plastic symbol templates in which each shape can be turned, shaded or half-shaded to

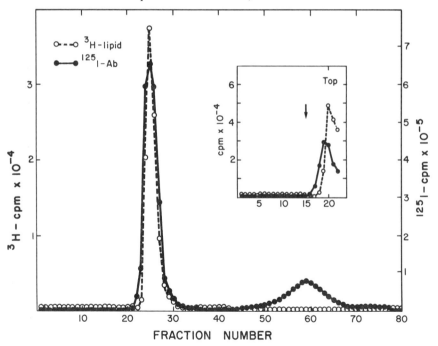

Figure 12.7. Combining two figures into one. (From Huang, Anthony, Leaf Huang, and Stephen J. Kennel. 1980. Monoclonal antibody covalently coupled with fatty acid. Journal of Biological Chemistry, vol. 255, No. 17, issue of 9/10. [pp. 8015-8018])

give a wide range of possibilities. If you are showing two or more different kinds of data, be sure that the symbols are also very different. If you want to give emphasis to one variable (maybe the control), make it stand out. One way would be to make that symbol solid and all the others hollow.

Charts

There are two kinds of charts: flow charts and pie or sector charts. A flow chart is used when you want to list options and indicate the order in which they should appear. Sometimes flow charts use words only, as in the example on page 50. Or words can be combined with a single drawing for a more interesting flow chart (Figure 12.8). Often scientific procedures are illustrated with a flow chart that uses no words at all (Figure 12.9).

Pie charts are used for the same sorts of data that bar graphs are, but are often more effective for showing percentage of

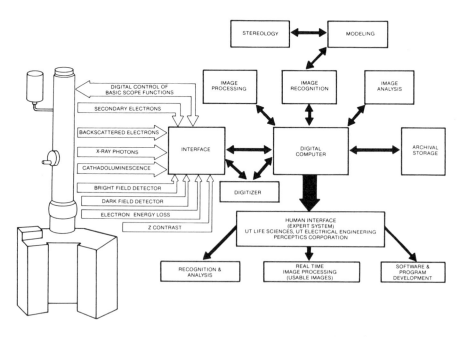

Figure 12.8. An example of a flow chart that combines words and drawings.

Pie charts are used for the same sorts of data that bar graphs are, but are often more effective for showing percentage of whole. Figure 12.10 illustrates the data on page 93.

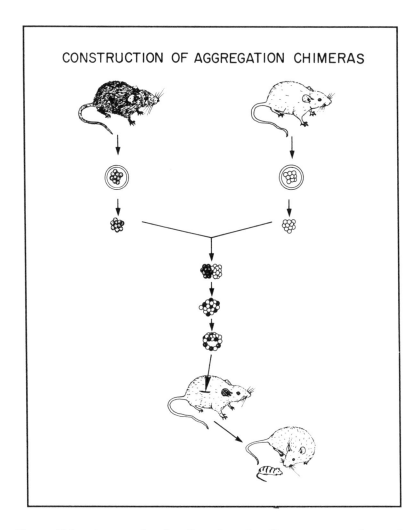

Figure 12.9. An example of a flow chart that illustrates a procedure, using no text. (Courtesy of Dr. Mary Ann Handel, from a poster session "Male Sterility of p^6H *and qk* is not rescued in mouse chimeras.")

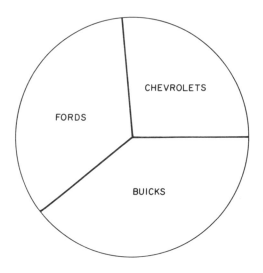

Figure 12.10. A pie chart based on the data shown on page 93.

The Final Draft

No matter how your figure will be reproduced, eventually your final draft will have to be black ink on white or translucent paper. You can then make good quality photocopies for a report, have photographic prints (called photostats, stats, PMTs, or glossies) made for publication and poster presentations, or have slides made. Even color slides with white lettering on colored backgrounds require the original artwork to be black on white.

If you are creating your figures on a computer, be sure that you are using a good quality printer and that your final copy has well-defined blacks and no gray smudges. Even if you have access to a computer, it is well worth the time to learn to produce your drawings by hand.

Paper

The easiest and most efficient method is to simply trace your rough draft after taping down a sheet of good quality tracing paper over it. Keuffel and Esser's Albanene and Teledyne Post's Blutex Vellum are both good and are available in most university bookstores. Bienfang's Satin Design is slightly more expensive but has the advantage that it can be erased many times; it can be found in large art supply stores. Drafting tape is best for taping down the tracing paper; if you can't find any, masking tape will work although it tends to be too sticky. Applying it several times to your shirt sleeve or jeans before you stick it to the paper will alleviate the problem.

Drawing Lines and Data Points

The easiest way to indicate lines and data points is to draw them with a technical pen which is designed to give an even and consistent ink flow. Keuffel & Easer and Kohinoor both make good pens in a large variety of widths. Both these companies also produce a good quality ink. Here is a checklist of other points to remember when drawing lines and symbols:

1. To make sure that the corners of your figure are square, use a T-square for the horizontal lines. For the vertical lines, use a 45°/90°/45° triangle on the top edge of the T-square.
2. Allow at least an inch margin. (The edges of the paper will not normally show but you may decide later to have slides made with color film.)
3. Be sure that your data lines are about 2 mm wide and are thicker than the axes and that no line is thinner than .5 mm.
4. The ink should look very black, not gray. If it looks gray, it may be an old bottle. Don't attempt to persevere with it — despite your best efforts, your figure may reproduce poorly. Start a new bottle. If the new one is also gray, take it back and ask for a replacement.

5. Probably the most prevalent problem in drafting is ink smearing under the T-square or triangle. Work from the top of your drawing down and remember that ink takes longer to dry in humid weather. Try using triangles with bevelled edges and hold the pen vertically, rather than at an angle. If your triangle does not have a bevelled edge, glue or tape a 1/2" strip of poster board under the ruling edges.
6. For curved lines, you can use either a flexible curve (which is easier) or a French curve (which requires a great deal of painstaking matching up of lines).

Using Press-on Lines and Symbols

A more expensive and less satisfactory way of indicating lines and symbols is to use press-on materials made by Zipatone, Letraset and others. Again, tape a sheet of translucent tracing paper over your rough draft and carefully apply the sticky backed lines where you need them (they are sometimes difficult to position straight). Symbols are simply rubbed on through the backing sheet. WARNING: If you choose this method, be sure to make a copy soon after completing the figure. The symbols occasionally fall off, especially if exposed to heat. Also, you will not be able to do tight curves or complicated map outlines with press-on lines.

Shading Large Areas

Often, especially on maps and bar graphs, some shading will enhance a figure's impact and overall appearance. If only a small area is required, you can draw diagonal lines (being sure to keep them evenly spaced) or stipple sparsely. If you need to shade a large area or are uncertain about your line drawing skill, shading films in many different patterns are available.

Some Points to Remember:

1. Always cut the film on the outside of the line outlining the area you wish to indicate. With age the film will shrink slightly. If you have cut it on the inside of the line, there will be an unsightly white area where the film has shrunk away from the line.
2. Select a film consistent with your meaning (wavy lines for water, tussocks for grassland, dots and lines for bar graphs, etc.)
3. If the film tears easily, refuses to lift off the backing sheet and seems very sticky, it is probably old. Don't try to persevere with it — the patched-together result is not worth the money you "saved." Also, don't try to stock up on extra shading film when it's on sale: it may be on sale because it's already old.

Lettering

All lettering on your figures should be clear and legible. Two methods are available: mechanical, using a scriber, technical pens and template ("Leroy" or "Wrico: lettering), and transfer ("rub-on") lettering. Although learning to use a Leroy set is sometimes difficult and time-consuming, it is a worthwhile skill to acquire (Figure 12.11). The templates are expensive but will last for years; they can also be borrowed. Additional advantages are that you will not be limited by the availability of rub-on letters, mechanical lettering (once mastered) is faster, and it is also less expensive since you don't have to keep buying sheets of lettters.

Some Points to Remember:

1. Use a T-square to rest the templates on (Figure 12. 11).
2. When spacing the letters, do it optically, remembering that an "i" needs more spaces than an "o."
3. When centering, you can find the middle of the word or phrase, and letter the front half backwards. This can be tricky so it may be worth the extra time to first letter the

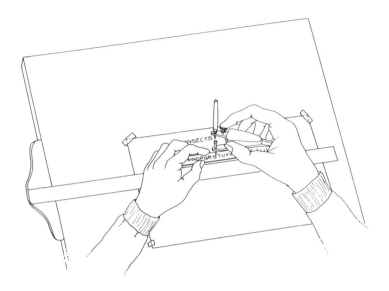

Figure 12.11. How to letter using a Leroy lettering set.

whole word on a piece of scrap paper, cut it out, arrange it where you want under the tracing paper, and trace it again. You can also tape the new word into the appropriate space, but sometimes it takes more time to splice in the new word than it does to re-letter it.

Although rub-on letters are expensive, they are readily available at large art supply stores and relatively easy to use. If you have used tracing paper over a rough draft drawn on graph paper, you will have little trouble applying the letters and numbers in a straight line.

Some Other Points:
1. Be sure to count how many numbers and letters you will need before buying the sheets. You can also buy sheets that contain only numbers.
2. Plan to use a typeface that is simple and easy to read. Helvetica Medium in a 14 to 16 point size is most generally used.
3. If the letters seem to crack, take the whole sheet back as it is too old to use.

4. Space the letters by eye. Remember to allow plenty of room between words.
5. If you need to center a phrase, find the middle, then letter backwards and forward from that center point. Spelling backwards is harder than you might think: having the word or phrase in front of you as a guide helps prevent disasters.
6. Have a copy or print made soon, as rub-on letters can fall off.
7. If you're careful, whole words can be re-positioned. Use a piece of 3M's Magic Mending Tape that is a couple of inches longer than the word and center it roughly over the letters. Rub it down firmly, leaving enough at one end to pick up. Grasp the end firmly and give it a good yank. The entire word or phrase should now be on the tape.

Another method of lettering which is sometimes available at copy shops is a Kroy machine. You can select a type size and style, insert the appropriate wheel and punch the letters you want which will appear on clear adhesive. Although convenient, this method can be quite expensive.

The most attractive method of lettering is to have the material typeset. This can also be very costly and is usually only worthwhile if the figures will appear in a book or very elaborate professional presentation.

Corrections

Doing your own illustrations for the first time can be difficult and sometimes frustrating. Fortunately, there are numerous salvage techniques available, depending on how your figures will be reproduced.

Photocopies and Prints. Photocopiers cannot "see" different shades of white. You can therefore correct mistakes by: (a) erasing; (b) using white-out (Liquid Paper dries quickly and has a good inking surface); (c) re-lettering on a piece of white artist's tape over the mistake; (d) splicing in a new piece of paper that includes the correction; (e) carefully scraping off excess ink with a very sharp razor or Exacto blade. A dull blade will only make holes in the paper.

Color Slides (Photographed with Color Film). Every kind of correction will show in this medium. If you must use color figures, do all the ink work first making whatever corrections you need to. Then have a print made — any corrections will no longer appear. Now you can apply colored shading film (Zipatone, etc.) to the print.

Color Slides (Hand-Colored). It is also possible to make your final drawing as usual on tracing paper with black ink. Have 35 mm negatives made with a high contrast film such as Kodalith, then carefully hand color the transparent areas using a waterbase marking pen ("Flair" pens work well) or watercolors. This technique works best for line graphs or other figures with small areas to be hand-colored.

Illustration as a Career Possibility

There are many people who started out as scientists and who then discovered that they got more joy out of illustrating research rather than doing it. Scientific illustration includes drawing plants, animals, machines, and artifacts as well as making maps and graphs.

For more information about illustration methods and careers, contact GNSI (Guild of Natural Science Illustrators), P.O. Box 77, Ben Franklin Station, Washington, DC 20044, or refer to the list of suggested readings.

Suggested Readings

Holmes, Nigel. 1984. *Designer's Guide to Creating Charts and Diagrams*. Watson-Guptill, NY.

Sonte, Bernard and Arthur Eckstein. 1983. *Preparing Art for Printing*. Van Nostrand Reinhold, New York.

Papp, Charles. 1976. *Manual of Scientific Illustration*. American Visual Aid Books. Sacramento, CA.

Wood, Phyllis. 1979. *Scientific Illustration: A Guide to Biological, Zoological, and Medical Rendering Techniques, Design, Printing and Display*. Van Nostrand Reinhold, NY.

Zweiful, Frances W. 1961. *A Handbook of Biological Illustrations*. University of Chicago Press, Chicago.

BIBLIOGRAPHY AND
SUGGESTED SOURCES

PHILOSOPHY OF SCIENCE

Ingle, J.D. 1958. *Principles of Research in Biology and Medicine.* J.B. Lippincott Co., Philadelphia.

Platt, J.R. 1964. Strong Inference. *Science* 146:347-353.

STATISTICS

Brown, F.F., et al. 1975. *Statistical Concepts, A Basic Program,* 2nd ed. Harper and Row, New York.

Guilford, J.P. 1978. *Fundamental Statistics in Psychology and Education,* 6th ed. McGraw-Hill Series in Psychology, McGraw-Hill Book Co., New York.

Rohlf, F.J. 1994. *Statistical Tables,* 3rd ed. W.H. Freeman Co., San Francisco.

Siegel, S. 1988. *Nonparametric Statistics for the Behavioral Sciences.* 2nd ed. McGraw-Hill Book Co., New York.

Sokal, R.R. 1995. *Biometry: The Principles and Practices of Statistics in Biological Research,* 3rd ed. Freeman, New York.

Steel, R.G.D. and J.H. Torrie. 1980. *Principles and Procedures of Statistics: A Biometrical Approach,* 2nd ed. McGraw-Hill Book Co., New York.

Terrace, H. and S. Parker. 1971. *Psychological Statistics.* 7 vols. Individual Learning Systems, Inc. San Rafael, California.

Zar, J.H. 1984. *Biostatistical Analysis,* 2nd ed. Prentice-Hall, Inc. Englewood Cliffs, New Jersey.

WRITING A SCIENTIFIC PAPER

Council of Biology Editors, Committee on Form and Style. 1994. *Scientific Style and Format: The CBE Manual for Authors, Editors, and Publishers.* 6th ed. Cambridge University Press, Cambridge, New York.

Gubanich, Alan. 1977. Writing the Scientific Paper in the Investigative Lab. *The American Biology Teacher.* January 1977.

Woodford, Peter F., ed. 1986. *Scientific Writing for Graduate Students: A Manual on the Teaching of Scientific Writing.* Council of Biology Editors, Bethesda, Maryland.

THE INVESTIGATORY LABORATORY

Biological Sciences Curriculum Study. 1976. *Research Problems in Biology, Investigations for Students.* Series 1, 2 and 3. 2nd ed. Oxford University Press, New York.

The Carolina Biological Supply Company publishes a newsletter, *Carolina Tips,* which is a good source of information on laboratory techniques.

Morholt, E. 1986. *A Sourcebook for the Biological Sciences.* 3rd ed. Harcourt Brace Janovich, New York.

(The appendix to this book has an excellent bibliography of sources for teaching biology and equipping laboratories.)

Thornton, John W., ed. 1972. *The Laboratory: A Place to Investigate.* Commission on Undergraduate Education in the Biological Sciences. Publ. No. 33. (Supported by a grant from the National Science Foundation to the American Institute of Biological Sciences.)

Turtox/Cambosco, MacMillan Science Co., a supply company for scientific equipment, publishes a newsletter, *Turtox News*, with useful laboratory hints and techniques.

APPENDIX

As indicated in the introduction to this handbook, students at any level can be prepared to perform original research. The essential prerequisite is a careful, gradual introduction to the scientific method. We offer you our general plan for the introductory weeks of any investigatory laboratory course not because it is the only one that works but because we have found it successful. We have frequently heard derogatory comments such as, "Only a few honors students can succeed in this sort of program," or "Non-majors are not capable of independent research," or "It's a fine idea but it won't work with freshmen," but we have never found them to be true. It *is* true that some inadequate educational systems manage to stifle creativity and independence, but, in the proper atmosphere they can be quickly revitalized and nourished. Students rarely exceed your expectations, so you must believe in them, encourage them, and guide them toward a scientific way of thinking.

In the introductory weeks outlined below, students learn to identify problems, ask appropriate questions about them, restate these questions in terms of null and alternative hypotheses, and design methods of answering these questions, figuring out what information would be needed and how to collect it.

I. First week in the laboratory.

At the first session, in his introductory remarks, the instructor points out that the primary methods of gathering scientific information are observation and experimentation. Before students learn to experimentally manipulate phenomena, they should consider the process of observation itself. It is pointed out that all knowledge is subjective, all information about the external world must be acquired by and filtered through the human nervous system. The best we, as scientists, can hope for is to learn to use our powers of observation

179

with as little human bias and subjective interpretation as possible. This is a matter of practice.

The students are then given specimens of plant and animal material. Typically these are leaves and crickets. (We usually have 3 habitat boxes, one containing 6 males, one with 6 females, and one with 3 of each sex, which the class can take turns examining.) The students are asked to relax (nothing will be graded), take their time, and make observations, writing down any data they consider significant. Later the data are collected, compared, and discussed as a group.

Most students collect good visual data but fewer record observations of sound, smell or taste. Many of their observations are colored by anthropomorphic assumptions. (For example, the female cricket is often assumed to be the male because the ovipositor is seen as a penis.) This should lead naturally into a discussion of bias and objectivity.

In the second laboratory period of the first week, students have more practice collecting data. The instructor drops two balls (tennis and ping pong) simultaneously in front of a vertical yardstick or meterstick, and the students, armed with tally counters and stopwatches, are asked to collect any data they feel are appropriate. The drop is repeated as often as necessary until the students are satisfied that they have completed their observations. Once again the data are collected, compared, and discussed. Students are not graded on their work.

In the discussion, it should quickly become apparent that different people collect different data because they have different questions in mind. Students may ask which ball bounces higher, or how many bounces does the ball make before it stops bouncing, or what is the average height of the first bounce, etc. The point to be underscored is that, in order to be meaningful, data must be aimed at answering a particular question.

After the discussion of the bouncing ball situation has subsided, the class should be urged to start asking questions about anything they can think of and then to restate these questions into answerable form. If no questions are forthcoming, students should mill about, look at various displays in the laboratory, look out the window, look at each other, etc. Any kind of question they might answer with simple observations or experimentation would be suitable for the purposes of discussion. Sample questions might be: What make of car passes the window of the lab most frequently? What is the average age (in months, perhaps) of the students in this class? How many of the students here are first children in their families? This exercise should lead to a discussion of the different kinds of data and different scales of measurement described in Chapter Two of this handbook.

Materials needed in the first week:

leaves

crickets (in habitat boxes if possible)

stopwatches

tally counters

two balls, visually distinct, and a meter or yardstick

II. Second week in the laboratory.

By the end of this week, students, working in pairs, should have performed four "mini-experiments" on simple systems such as dice, cards, spinners, coins, balls and metersticks, or other phenomena of their choice. They should be able to come up with questions, refine the questions into answerable form, and restate them in terms of null and alternative hypotheses. In addition, they should be able to determine what kind of data would answer their questions, how much data would be needed, and which statistical test would be appropriate to determine the statistical significance of their results. Naturally, at the beginning, the students will need a lot of help. Class discussions about the various questions and data collected are particularly helpful.

At the opening laboratory session of week two, there should be a short review of asking questions. This time the questions should be restated in terms of null and alternative hypotheses. As pairs of students find questions they feel they would like to use for their mini-experiments, they should be encouraged to begin. Research pairs should start by writing a brief outline of their research proposal. This should contain the question being asked, the null and alternative hypotheses, an outline of the methods and materials, and mention of the statistical treatment to be applied to the results. These written proposals should be checked by the instructor before the team begins its data collection.

If students are having trouble finding questions that differ from examples already used by the instructor, they should be encouraged to think of comparative questions. For example, if all they can think of when looking at a coin is to ask whether or not the coin is fair (an example they would have gotten from the handbook), they should be moved toward questions such as the following: Can Charlie flip more heads than Sue? Do the results of tosses caught on the hand differ from those of tosses which land on the floor?

The students should write brief reports of the results of their experiments, including the actual data collected, its analysis, and its statistical significance. Mini-experiments should take approximately one hour from proposal to report. These experiments should be graded in terms of clarity of thought and experimental design. It is not important that the results be statistically significant, although conclusions drawn should be consistent with the original question asked and the actual data collected.

Materials needed in the second week:

stopwatches

tally counters

coins, dice, spinners, playing cards or other
   gambling paraphenalia
balls and metersticks

III. Third week in the laboratory.

This week, working in pairs, the students design and perform an experiment of their choosing concerning the human blood circulatory system. Usually the experimental questions concern the rate of heartbeat, the blood pressure, or the body temperature (or some combination of these) because these are aspects of the system that are easily measured.

Fellow students and any others they can persuade to cooperate are available as experimental subjects. Frequently faculty, secretaries, janitors and other staff are railroaded into volunteering as experimental subjects.

The research pairs have one week to design their experiment, have their proposal accepted by the instructor (instructors should be careful to consider the safety of the project), gather their data, analyze it, and write it up. Preparation, data collection, analysis, and writing usually take place in the laboratory.

Materials needed for the third week:

stethoscopes

sphygmomanometers

thermometers

stopwatches

people willing to be experimental subjects

handouts of background information

(We have prepared a description of the human heart and related phenomena such as pulse and blood pressure, mentioning the roles of hormones and the sympathetic and parasympathetic nervous systems. We also have several "how to" sheets: How to measure the rate of heartbeat. How to measure body temperature. How to measure blood pressure.)

IV. Fourth week in the laboratory.

This week, working in pairs, the students design and perform an experiment relating to germination in the seeds of the alfalfa plant, *Medicago sativa*. Once again the students must come up with their own questions, formulate their hypotheses, and have research proposals approved by their instructors. Most phases of this study are carried out in the lab. A written research report is required and is usually prepared outside class.

Materials needed in the fourth week:

seeds of alfalfa, *Medicago sativa*

Petri dishes

paper toweling (for providing a suitable, damp environment for germination)

tissue paper layers or aluminum foil (to cover Petri dishes with materials allowing differing degrees of light penetration)

thermometers (for locating and recording different temperature environments, for example)

potting soil or sand (for varying the depth of germination and other phenomena associated with depth such as oxygen availability)

salt, fertilizer, dextrose, sucrose, glucose, and possibly other substances (to vary the growth environment in terms of its acidity, nutrients, etc.)

rulers (to measure root production as an indication of germination)

temperature controlling devices such as refrigerators, heaters, etc.

handouts discussing dormancy and germination and suggesting possible environmental factors which might make seeds come out of dormancy.

V. Fifth week in the laboratory.

The students have one more week-long experiment to perform and write up. This time there is much more freedom of choice in the topic. They are asked to inspect the laboratory supplies such as Petri dishes, glassware, plastic boxes, lamps, hot plates, ice, aluminum foil, colored paper and cellophane, stopwatches, thermometers, respirometers, sand, soil, and various chemicals, making a mental inventory of what is available. Then they are given an organism list of specimens which the staff has on hand for student use. A typical list might be: pill bugs, roaches, planaria, snails, flour beetles, crickets, meal worms, seeds, and algae. Research teams select their organism and think up any experiment they can using the lab supplies or things they manufacture themselves. Again, research proposals must be approved and final reports submitted.

A suggestion sheet is usually provided to help them find suitable research questions. Typical topics from the suggestion sheet are:

Habitat selection. Do some animals actively seek out certain temperatures, pH, light conditions, colors, etc?

Distribution. How do the animals distribute themselves in their environment? Are they in clumps, randomly distributed, or uniformly distributed? What factors seem to influence the distribution process?

Orientation. What directional cues are the animals using as they move about?

Social behavior. (Particularly suitable for crickets) Are the crickets territorial? Do they have a social hierarchy, and, if so, what factors seem to influence social rank?

Respiration. How do different environmental factors (such as temperature, light, etc.) affect the behavior (or germination) of the organism? A handout is available on how to determine the rate of respiration for air-breathing animals and seeds.

VI. From week six to the end of the term.

During the final weeks of the term, the students design, execute, and write up one major scientific experiment. This time the topic is geared to mesh with the material being studied in the course. For example, if the students are learning genetics, they might do anything from a classical *Drosophila* experiment to a survey at shopping centers to see if human populations are in a Hardy-Weinberg equilibrium. In other quarters or semesters of the course, the research projects are also coordinated with the broad areas being studied.

For the major projects, lists of possible research topics and organisms are provided and a few handouts with techniques and background information. This time, in preparing the full (no longer in an outline) research proposal, students must search the literature in the library to inform themselves about their topics and locate published techniques where needed.

Research proposals must be carefully examined to be sure that the students understand their questions and how they will answer them. Both grandiose and overly simple experiments should be discouraged so that no student is working either above or below his capability.

Research results should be reported in the form of a full, formal, scientific paper with the necessary figures and tables carefully prepared. Once again, the grading should consider the experimental logic, clarity of presentation, accurate analysis of data, and intelligent conclusions based upon results. Whether or not the experiments actually succeed in providing a statistically significant answer to the question posed should be inconsequential. In all cases, the scientific approach used, not the actual "success" of the experiment, is paramount. Needless to say, the research performed does not have to be original as long as it results from a question which is genuinely original to the student. The work will be valuable whether it finds something previously unknown or merely supports other similar findings in the field.

VII. Why we teach this way.

By this method of teaching, we hope to give each student the ability to think critically and analytically, to ask pertinent questions, and to design tests that would answer them. This way of thinking and method of problem solving should foster both curiosity and ingenuity and should give each student a taste of the excitement of science. The Investigatory Laboratory method of teaching is a truly self-paced, self-directed, and particularly self-rewarding experience. Once a student is armed with the scientific method, he can discover and uncover facts for himself — he holds the key to unlock the secrets of the universe.

GLOSSARY/INDEX

abscissa — The abscissa is the horizontal axis of a graph. In a graphic representation of data such as a frequency histogram, the vertical side, or ordinate, designates the frequency of the values recorded and the abscissa is used for the different scores or values of the parameter being measured. The lowest value occurs at the left end of the abscissa. p. 14.

abstract — An abstract is a one or two paragraph condensation of the main elements in a scientific article. For sample abstracts from published scientific papers, see p. 113-115. Because an abstract can give the reader a quick impression of the material published, abstracts are often published separately by services which provide a guide to current literature. See p. 104-106.

alpha — The "alpha level" (or α) is an arbitrary level of risk. When using a test to determine whether or not you reject your null hypothesis, you knowingly set the level of risk, such as a 5% chance, that your test will lead you to accidentally reject a true null hypothesis (a Type I error). p. 46.

ANOVA — ANOVA, or the analysis of variance, is a parametric statistical test of differences between means of more than two samples. p. 51.

asymptotic — This term is used to describe the relationship between the tails of a normal curve and the base line (abscissa) of a graph. The tails of the curve of a normal distribution are asymptotic to the base line, or abscissa, receding indefinitely and never actually touching the baseline because the perfect normal curve represents an infinite number of cases. p. 24.

bimodal distribution — When a group of ordered data, plotted on a frequency histogram, reveals two peaks of frequently occurring values, it is described as having a bimodal distribution. p. 18.

Chi Square — see X^2.

conclusion — For a discussion of drawing conclusions based on the results of statistical tests, see p. 47-48. In writing a scientific paper, the conclusions you draw from your experimental results are presented in the section entitled "discussion." For sample discussion sections, see p. 128-133.

continuous data — Continuous data are variables from a scale such as length or time which is a continuum. Although there will be actual gaps between the items measured and recorded, there is always the theoretical possibility of additional data points between them. Unlike the measurement of numbers of children, discrete data with no possible child-and-a-half fractions falling between two consecutive readings, in continuous data fractional intermediate values are always possible. p. 10.

correlation — Variable factors which are related to each other, such as shoe size and height are said to have a correlation. Correlation does not necessarily imply a cause and effect relationship between the two variables. In the example mentioned, since taller people usually have larger shoe sizes, the factors are said to have a *positive* correlation. Altitude and air pressure have a *negative* correlation because, as altitude increases, air pressure decreases. p. 39.

discrete data — Discrete data consist of items that are separate, whole units. The numbers of children in sample families would be plotted as discrete, whole number units. There cannot be a fractional increment on the scale between 1 child and 2, or between 2 and 3 children, etc. p. 8.

discussion — In a formal, scientific publication, the discussion is the last section of the text before the "literature cited." The discussion normally presents the conclusions drawn from experimental results and analyzes these findings in terms of the question posed and the current state of knowledge about the topic under consideration. For samples, see p. 128-133.

dispersion — The dispersion of data is its spread on either side of a central tendency. Dispersion can be measured in terms of its *range* (the distance between the highest and lowest reading) or its *variance*, a measure of how clumped the data are. p. 22-29.

distribution — see frequency distribution.

F Test — The F test is a test for differences between variances. p. 85-87.

frequency distribution — When data such as scores, or other measurements of some variable population, are arranged in order of magnitude, this distribution reveals the frequency of occurrence of different values. When the ordered scores are represented in a table or graph which indicates the number of times a particular score or value occurs in the group, the data are represented in a frequency distribution. Frequency distributions of data are often displayed in a frequency histogram. p. 14.

frequency histogram — The frequency histogram is a graphic representation of data indicating how frequently different values occur. In the graph, the frequency of occurrence is plotted on the vertical axis (ordinate) with frequency increasing with the height of the axis. The data being plotted, whether discrete or continuous, are distributed along the horizontal axis (abscissa). In the case of data which can be ordered, the direction of increase is from left to right. In a histogram, the scores are represented by rectangular boxes of differing heights (frequencies) over the appropriate points on the abscissa. See p. 14.

Friedman Test — The Friedman Test is a non-parametric test of differences between means; it is used to test for significant differences between the responses of several matched samples (paired samples) exposed to three or more treatments. p. 58-62.

hypothesis — An hypothesis is a possible explanation for a phenomenon. Scientific investigation involves the testing of predictions based on hypotheses. (See also null hypothesis.) p. 1.

interval scale — There are two scales of measurement for continuous data — the interval and the ratio scale. Both scales of measurement indicate the distance between items of continuous data, but the units of measurement on the interval scale are not fixed by an absolute zero point. Zero degrees Fahrenheit is an arbitrary endpoint which does not signify the absolute absence of heat. January 1 is an arbitrary starting point but can hardly be considered the beginning of time. p. 10.

introduction — In a scientific paper, the introduction presents the question being asked, or hypothesis being tested, and justifies the experiment by indicating what is currently known about the topic under investigation and why this particular question is of interest. Findings by other authors are usually mentioned in this part of the paper. For examples of introductions, see p. 115-119.

Kruskal-Wallis — The Kruskal-Wallis Test is a non-parametric test of differences between means. p. 63-67.

linear regression — See regression analysis.

literature cited — The literature cited is the last formal section of a scientific paper. It is a bibliographic list of published material actually mentioned in the article and is presented in alphabetical order based on the surname of the first author of each entry. For a brief discussion see p. 134. For help in deciding what bibliographic data to collect when doing a literature search, see p. 102.

mean — The mean (often written as $\bar{X}$), a calculation of central tendency in a frequency distribution, is the numerical average. The mean is computed by dividing the sum of the scores by the number of scores. p. 20-21.

median — The median, a measure of central tendency, is the middlemost value in a distribution of data. To find the median, the data scores or values recorded must be organized in a progressive sequence. In the case of an even number of readings, an exact median can be calculated by averaging the two central data points (adding them together and dividing the sum by two). p. 20-21.

methods and materials — In a formal scientific paper, the section entitled methods, or methods and materials, describes the materials and procedures used in the investigation being reported in sufficient detail that another scientist could repeat the experiment. Previously published techniques can be cited and not repeated in detail. Even though this section is supplied so that an experiment could be repeated for purposes of verification, it should be written in the past tense, telling what was done not what the reader should do. See p. 119-124 for examples.

mode — The mode, a measurement of central tendency, is the most frequently occuring value in a distribution of data. It is possible for a distribution to have more than one mode. The mode is easily revealed if the data points are ordered sequentially or plotted on a frequency histogram. See p. 20-21.

nominal scale — Discrete data, "measured" on a nominal scale, are data assorted into named groups. Nominal measurement usually involves counting items in the separate categories. (See ranking scale for comparison.) p. 8-9.

normal distribution — Large groups of quantitative data usually approximate the theoretically normal frequency distribution characterized by a symmetrical, "bell-shaped" curve with a mean, median, and mode of the same value. This curve, a frequency histogram smoothed of its steps because all theoretical points are illustrated, has an equal number of scores on either side of the central axis (or mean). The tails of the curve are asymptotic to the baseline (abscissa) infinitely receeding and approaching but never touching it. There are two points on the normal curve where the direction of curve changes from convex to concave. These are called the points of inflection. See pages 23-27 for more information.

null hypothesis — The null hypothesis (H_o) is the hypothesis of no difference. All statistical tests are designed to determine whether or not you have sufficient reason to reject your null hypothesis. Since there is no such thing as positive proof, scientific advances are made by the rejection of null hypotheses. If your data indicate a statistically significant reason to reject your null hypothesis, for example that the mean of sample A is equal to the mean of sample B (let us say at the .05 alpha level of confidence that you have not made a mistake), then you can accept one of your alternative hypotheses (that the mean of A is greater than or less than the mean of B). This type of logical basis for a statistical test is typically written in shorthand:
$$H_o : \bar{X}A = \bar{X}B \quad H_1 : \bar{X}A > \bar{X}B \quad \text{or} \quad H_2 : \bar{X}A < \bar{X}B. \text{ p. 32-34.}$$

majority of the data you collect will be measured on a ratio scale where the possibility of no growth, no time elapsed, no distance traveled, etc. will exist. p. 10.

regression analysis — regression analysis deals with the study of dependent relationships between correlated variables. There is a brief discussion of this kind of correlation on p. 40. The statistical test for simple linear regression begins on p. 72.

skewed distribution — When data plotted on a frequency histogram are not evenly distributed on either side of a central high point, they are skewed. In positively skewed data, the mean is located to the right of the majority of the readings. In a negatively skewed distribution, the mean is located to the left of most of the values plotted. p. 16.

Sign Test — The Sign Test is a non-parametric test of differences between means. p. 78-80.

Spearman's Rank Correlation — The Spearman's Rank Correlation is a non-parametric test of relationships which may be used to determine whether or not two variables are correlated. p. 81-84.

standard deviation — The standard deviation (written as "s" or σ), is a standard unit of measurement of deviation or distance from the mean along the abscissa of a frequency distribution. This unit is a measurement away from the mean point on the baseline (to the left and/or to the right) to the point where a theoretical, perpendicular line from the point of inflection crosses the abscissa. Because the tails of the normal curve are asymptotic to the baseline, an infinite number of standard deviations from the mean exist, but three standard deviations from the mean, on either side (a total of 6), account for 99.74% of all the values, *by definition*. One standard deviation on either side of the mean (a total of 2) includes 68.2% of the possible data values, and two standard deviations (a total of 4) include 95.44%. p. 24-25.

t Test — The t Test is a parametric test for differences between means of independent samples. p. 85-91.

Type I and Type II errors — When using a statistical test to evaluate a null hypothesis, there are two types of errors you can make. *A Type I error is the rejection of a true null hypothesis.* If you set alpha at the .05 level of confidence, the chances are 1 out of 20 that your statistical test will accidentally allow you to reject a null hypothesis that actually was true. *The Type II error is the failure to reject a false null hypothesis.* A lower alpha level will decrease your chance of making a Type I error but will increase the chance that you will fail to reject a null hypothesis that really is false. (The .05 alpha level is the normal compromise position between these two risks). p. 45.

ordinate — The vertical axis of a graphic presentation of data. In a frequency histogram, the ordinate indicates the frequency of occurrence of the values measured, with frequency increasing with the height of the ordinate. p. 14.

ordinal scale — An ordinal scale, or ranking scale, of measurement is used to organize items of discrete data in terms of some meaningful relationship to each other. The nominal categories can be ranked according to some aspect. For example, categories of makes of cars can be organized in terms of their increasing weight along an ordinal or ranking scale (and along an abscissa). p. 8-9.

parametric statistics — Parametric statistical tests are limited to statistically normal distributions of data whose data points or individual observations are independent of each other and distributed on the same, *continuous* scale of measurement. In general, if these criteria can be met, parametric statistics should be selected over non-parametric tests because they are more "powerful" tests, giving you a greater ability to reject your null hypothesis. p. 44.

parameters — Parameters are aspects of a population of data such as the mean or the range. p. 8.

population — The term population is used in statistical tests to describe any group of similar kinds of things being measured. p. 8.

points of inflection — In the bell-shaped curve characteristic of a normal frequency distribution, the points of inflection are those points on either side of the mean where the direction of the curve changes from convex to concave. Theoretical lines drawn from the points of inflection to the abscissa mark off the standard deviation on either side of the mean. p. 24.

range — The range of a distribution is a measurement of its dispersion around the mean. It is the distance between the lowest and the highest reading. p. 22.

Rank Sum Test — The Rank Sum Test is a non-parametric test of differences between means. p. 68-71.

ranking data — To rank a series of values, you assign the rank of 1 to the lowest, 2 to the next highest, and so forth. See the illustration in the Spearman's Rank Correlation Test on p. 81.

ranking scale — The ranking scale is the same as the ordinal scale of measurement. It is a meaningful order or ranking imposed on categories of discrete, nominal data. p. 8.

ratio scale — The ratio scale is a scale of measurement for continuous data. Unlike the interval scale, the ratio scale has a true zero point although the units of measurement are arbitrary. The

uniform distribution — In a uniform distribution of data, the frequency of occurrence of each value is the same and the bell-shaped normal curve is flattened to a straight line. p. 15.

variance — The variance is a measurement for describing the dispersion of data around the mean. By definition, it is the square of the standard deviation. It is written as s^2. Although it is a useful parameter for certain statistical tests, it is not plotted on most displays of data where the standard deviation should be used because of its size. p. 23 and 29.

X^2 — The X^2 (Chi Square) One-Sample Test for Goodness of Fit is a test for differences between distributions. p. 92-94. The X^2 Test of Independence between two or more samples is a test for independence between two or more frequency distributions of nominal data. p. 96-98.